AF588334

Plant Biotechnology for Health

María Alejandra Alvarez

Plant Biotechnology for Health

From Secondary Metabolites to Molecular Farming

María Alejandra Alvarez
School of Pharmacy and Biochemistry
CONICET/Universidad Maimónides
Buenos Aires, Argentina

ISBN 978-3-319-05770-5 ISBN 978-3-319-05771-2 (eBook)
DOI 10.1007/978-3-319-05771-2
Springer Cham Heidelberg New York Dordrecht London

Library of Congress Control Number: 2014944314

Printed on acid-free paper

Springer is part of Springer Science+Business Media (www.springer.com)

Preface

In vitro plant cell cultures have provided a tool for studying plant metabolism and physiology and to explore productive processes of secondary metabolites.

More recently, genetic engineering has allowed the use of *in vitro* cultures to modulate plant biosynthetic pathways and to express heterologous proteins of biomedical relevance.

The aim of this book is to provide some background information about the potential of *in vitro* cultures for conducting phytofermentations and for molecular farming.

Also, the application of *in vitro* cultures for the production of the steroidal glycoalkaloid solasodine by *Solanum eleagnifolium* Cav. and the catalytic antibody 14D9 is reviewed. The expression of a recombinant glycoprotein E2 from BVDV in *N. tabacum* is reported.

I hope this book drive attention to the potentiality of *in vitro* plant cultures as a supply of compounds to be used for diagnostics and therapeutics.

I would like to thank Dr. Patricia Marconi and Dr. S. Klykov for their collaboration in the chapter related to mathematical modeling in recombinant plant systems. Also, I would like to thank the School of Pharmacy and Biochemistry of Universidad Maimónides and the National Agency of Promotion of Science and Technology (ANPCYT), from Argentina for their support.

Buenos Aires, Argentina
2014

María Alejandra Alvarez

Contents

About the Author

María Alejandra Alvarez completed her studies at Universidad de Buenos Aires, Argentina. She graduated in Biochemistry in 1981 and Pharmacy in 1994 and obtained an MSc in Industrial Microbiology and Biotechnology in 1986 and a PhD in Plant Biotechnology in 1993.

She was lecturer in graduate and postgraduate courses of Botany, Pharmacobotany, Biotechnology, Food Biotechnology, and Galenic Pharmacy at Universidad de Buenos Aires, Universidad Nacional de La Plata, and Universidad Maimónides.

Dr. Alvarez started working on the production of compounds of pharmaceutical interest in *in vitro* plant cultures in 1986, being the author of numerous scientific articles. Her research interests include chemotaxonomy, medicinal plants, production of secondary metabolites, and molecular farming.

She was a researcher at Universidad de Buenos Aires and Centro de Ciencia y Tecnología Dr. César Milstein (CONICET/Fundación Pablo Cassará). Currently, she is a member of the National Council of Research and Technology (CONICET) in Argentina, Professor of Pharmacobotany and Pharmacognoscy and Director of the Plant Biotechnology Group at Universidad Maimónides, and Professor of Biology at Colegio Divino Corazón.

List of Figures

List of Tables

Chapter 1
Plants for Health: From Secondary Metabolites to Molecular Farming

There is a continuous search of new drugs and molecules for health care. Plants offer a huge variety of those bioactive molecules, some of which are still unknown. Popular medicine is the source of information for ethnobotanists and pharmacobotanists, which in general are the first link in the chain that leads to the taxonomical, chemical and pharmacological description of a species.

Chapter 2 summarizes the classical strategy employed to analyze the occurrence of bioactive principles in plants, since their collection in the field to the analytical and biological assays. It also refers to the contribution of new technologies (e.g.: systems biology, synthetic biology) to the knowledge about medicinal plants, their active principles, and their specific contribution in the treatment of some pathologies.

Chapter 3 exposes the classification and main characteristics of plant secondary metabolites, giving some examples of their applications for health. The pharmaceutical industry profits from those bioactive molecules to elaborate or, taking plant compounds as starting point or model, to synthesize drugs with known or novel pharmacological activities.

The traditional sources of raw material are plants in the field. However, in the second half of the twentieth century the *in vitro* culture technology arise bringing the possibility of maintaining plant cultures in controlled environmental conditions, free from the risks of pathogen and predator attack, from weather variations, and from geopolitical circumstances. Chapter 4 describes the main characteristics and prospectives of *in vitro* cultures.

Chapter 5 reports the strategy employed to analyze the production of a solasodine, a steroidal glycoalkaloid, as an example of the approaches used to develop a productive process of a secondary metabolite in *in vitro* cultures.

By the end of the twentieth century the production of recombinant proteins in plants, named molecular farming, appeared as a new strategy for obtaining biopharmaceuticals. Numerous proteins were expressed (antibodies, antigens, blood proteins, growth factors, etc.) but, in most of the cases, the attained yields were not competitive to the traditional production platforms. There still more work to do before establishing a competitive production platform of plant-made biopharmaceuticals.

M.A. Alvarez, *Plant Biotechnology for Health: From Secondary Metabolites to Molecular Farming*, DOI 10.1007/978-3-319-05771-2_1

However, several companies are developing this technology and some products are in the pipeline or already in the market. Chapter 6 describes the basics of molecular farming and some of their applications.

Chapter 7 is about the expression of a whole recombinant antibody in plants, the catalytic antibody 14D9, which was taken as an experimental model to produce a recombinant protein in *in vitro* cultures. The expression of 14D9 was accomplished in undifferentiated *in vitro* cultures (calli and suspended cells), in hairy root cultures and a preliminary scale up to a 2-l bioreactor was performed.

Chapter 8 refers to the potential application of a recombinant immunogen, the truncated glycoprotein E2, transiently expressed in plants, to trigger a specific immune response and eventually to be used to formulate an experimental vaccine.

Finally, Chap. 9 gives a general view of mathematical modeling for producing recombinant proteins in *in vitro* cultures under Good Manufacturing Practices.

Chapter 2
Plants for Health

Abstract Humankind has been always employing plants as supply of food, textile, cosmetics, medicines, etc. Their chemical compounds have biological activities that are effective for the treatment of different human and animal diseases. So far, on an estimated number of 250,000–300,000 plant species only approximately 5,000 have been intensively studied for their medical properties. Ethnobotany and Bioprospecting can contribute to identify new plant species with potential pharmacological properties. When a new species is identified as potentially useful for health, two main approaches are practicable, one aiming to the search of an already existing biological activity that is in the market, the other to the random screening of any biological activity. Pharmacobotany, Pharmacognoscy and Phytochemistry, through a classical approach, contribute to identify chemicals and pharmacological activities by screening plant species using different fractioning methodologies, chemical assays and bioassays (*in vitro*, *in vivo*, *ex vivo*). New strategies as *in silico* predictions, which gather information about drug-target interactions and human diseases, System Pharmacology and Polypharmacology can increase the current knowledge on medicinal plants through novel approaches. A multidisciplinary work that integrates classical with new strategies will gradually have more impact in the identification of plant bioactive principles and their potential uses in health care.

Keywords Medicinal plants • Screening of plant active principles

2.1 Introduction

Plants are being used as source of food (cereals, legumes, vegetables, fruits, etc.), animal feeding (clover, millet, alfalfa, etc.), and essences for the elaboration of perfumes and flavors (rosemary, sage, lavender, mint, etc.). Also, plant resins and gums are employed in tanneries, woody species for construction and carpentry (oak, cedar, ebony, pine) and for the production of cellulose and paper (poplar, pine, fir). Other plant species produce textile fibers (linen, cotton and hemp) or are appreciated as ornamentals, e.g.: roses, orchids, tulips, dahlias, etc. (Table 2.1). Plants have also been used for obtaining medicines (e.g.: digital, aconite, rhubarb, belladonna, henbane, etc.), and people have been transferring the knowledge about plant properties and their employment from one generation to the next one (Teiten et al. 2013;

M.A. Alvarez, *Plant Biotechnology for Health: From Secondary Metabolites to Molecular Farming*, DOI 10.1007/978-3-319-05771-2_2

Table 2.1 Commercial uses of some plant species

Industry	Species	Use
Food	*Coffea arabiga*	Beverage
	C. sinensis var. *sinensis C. sinensis* var. *Assamica*	Flavor
	Vanilla planifolia	
	V. tahitiensis	
	Piper spp. (Piperaceae)	
Cosmetic	*Lythospermum erithroryzum*	Dye
	Lavandula spp.	Perfume
	Mentha spp.	Flavor
Pharmaceutical	*Atropa belladona*	Cardiotonic
	Digitalis lanata	
Textile	*Gossipium hirsutum*	Fabric
	Linus usitattissimum	
Agrochemical	*Chrysanthemum cinenariaefolium*	Insecticide
	Calceolaria andina	
Construction and Carpentry	*Schinopsis balansae*	Wood
	S. lorentzii	
	Quercus rober	
	Cedrus spp.	
	Pinus spp.	

Verpoorte et al. 2006). In the last centuries, the pharmaceutical industry has profited from the uncountable varieties of plant bioactive molecules to produce medicines and has synthesized molecules of similar biological activities. Aspirin was the first synthetic drug (1897), and was the starting point for the development of methods and technologies for synthesizing plant active principles or related compounds (Schmidt et al. 2008).

In Western countries, where chemistry is the backbone of the pharmaceutical industry, 25 % of the molecules used are of natural plant origin (Payne et al. 1991). On the other hand, the World Health Organization has estimated that more than 80 % of the world's population in developing countries depends primarily on herbal medicine for basic healthcare needs (Vines 2004).

2.2 Active Compounds from Plants

After a species is recognized as potentially useful for the pharmaceutical industry, their active principles and properties are studied. In some circumstances the crude plant extract or the pure active principle is directly employed as medicine. If the active chemical principles cannot be isolated or there is a shortage on raw material, the pharmaceutical industry focuses their efforts on producing that molecule by synthesis or semi-synthesis (Raskin et al. 2002; Balunas and Kinghorn 2005; Doughari 2012; Moronkola 2012).

Table 2.2 Active pharmaceutical compounds from plants

Active principle	Source	Use
Artemisinina	*Artemisia anua*	Anti-malarial
Atropine, Hyosciamine, Scopolamine	*Atropa belladonna*	Anti-cholinergic
	Datura spp.	
Capsaicine	*Capsicum* spp.	Analgesic
Codeine, morphine	*Papaver somniferum*	Analgesic, Anti-tussive
Cocaine	*Erytrhoxylum coca*	Local analgesic
Cyanogenic glycosides	*Manihot esculenta*	
Digoxin, digitoxin	*Digitalis* spp.	Cardiotonic
Diosgenin	*Dioscorea* spp.	Steroidal contraceptive
Genistein	*Glycina max*	Anticancer
Nicotine	*Nicotiana* spp.	Anticancer
Pilocarpine	*Pilocarpus jaborandi* L.	Cholinergic
Psolaren, Xhanothotoxin	*Apium graveolens*	Anticoagulant
Quinine	*Cinchona* spp.	Anti-malarial
Reserpine	*Rauwolfia serpentine* L.	Anti-hypertensive, psychotropic
Taxol	*Taxum brevifolia*	Anti-neoplasic
Vinblastine, Vincristine	*Catharantus roseus* L.	Anti-neoplasic

At the end of the twentieth century, from the new 3,500 novel chemical structures identified, 2,619 were isolated from superior plants. In the United States of America, 25 % of the market is based on plant-based compounds. What is more, most of the prescribed drugs of plant origin were initially used in popular medicine. Table 2.2 summarizes some plant-derived compounds of pharmaceutical interest.

Up to now, only 5,000 of plant species over an estimated number of 250,000–300,000 were intensively studied for their medical properties. Unfortunately, a great amount of them are in risk of extinction (Payne et al. 1991).

Ethonobotany and Bioprospecting have the potential to contribute to the discovery and conservation of plant sources of bioactive molecules.

2.3 Ethnobotany

The term Ethnobotany was coined in 1895 as “the study of plants used by primitive and aboriginal people” by the botanist Harsherberg. The term was extended years before to “the study of the interrelationships of primitive men and plants” (Jones 1941). Currently, it is defined as “the study of the relationships established between people and plants” (Albuquerque 1997) without specifying a particular time frame. What is for sure is that the information about the properties of plant species of a particular region gathered by ethnobotanists may be valuable to modern medicine. Those species can thus be used directly as a source of a medicine, as raw material for the semi-synthesis of bioactive compounds, as a model for the synthesis of new molecules, or as taxonomic markers for the discovery of new compounds (Gurib-Fakim 2006). In ancient times, naturalists, travelers, and missionaries collected information about habits and practices of people in distant and new discovered communities or tribes, which was often deposited in missions, monasteries, abbeys, etc. Ethnobotanists look in those documents and in archeological sites for information about the connections between ancient civilizations and towns with their plants. Those data reveal the migration habits, commercial routes, agricultural practices, and the origin and dispersion of those plant species. Ethnobotanists also study the relationship between people and plant species of today compiling the knowledge guarded by indigenous and rural people.

In Argentina, ethnobotany dates from the eighteenth century, when the Jesuits in the Missions collected information from the natives. The Plantae Diaphoricae Florae Argentinae (Hieronymus 1882) is considered as the first compilation of Argentinean plants in use. Domingo Parodi, Eduardo Matoso, Nicolás Rojas Acosta, Juan A. Domínguez, Lorenzo R. Parodi, Arturo Ragonese, and N. Martínez Crovetto made significant contributions to the matter (Pirondo and Keller 2012). Currently, numerous ethnobotanical studies were performed in different regions and communities of Argentina (Zardini 1984; Arenas 1997, 2012; Molares et al. 2009; Rosso 2013; Scarpa and Montani 2011; Suárez 2009).

2.4 Bioprospecting

Bioprospecting, or biodiversity prospecting, is the exploration of biological materials for commercially valuable genetic and biochemical properties. Local communities provide the information about collection and properties of those potentially valuable resources for the pharmaceutical, biotechnological, and agri-business fields. Bioprospecting supporters consider it as a means of preserving biodiversity, by financing its conservation, and of discovering new drugs for human health (Harvey and Gericke 2011). For supporting the sustainable use of biodiversity the United Nations Convention on Biological Diversity was created in Rio de Janeiro (1992) with the objective of preserving biodiversity, promoting the sustainable use of the components of biodiversity, and encouraging the sharing of benefits derived from the commercial and other uses of genetic biological resources in a fair and equitable way. In the USA, the NIH funded the International Cooperative Biodiversity Groups (ICBG) that coordinates cooperation programs involving the USA and countries with high biodiversity as Costa Rica, Indonesia, Panama, Papua New Guinea, Madagascar, the Philippines, and Vietnam. With the same goal, the Rutgers University funded the Global Institute for BioExploration, GIBEX. The Bonn Guidelines and the Nagoya Protocol also considered the non-monetary benefits related to the commercialization of natural resources (Castree 2003).

Usually, companies establish a financial agreement with countries with high biodiversity (e.g.: Costa Rica, Brazil), which is a point that leads to controversies about the inappropriate commercialization and to the approach used to valuate biodiversity

(Harvey and Gericke 2011). Detractors consider bioprospecting as "free market environmentalism" or as a "post modern ecological capitalism" highlighting the unfair distribution of benefits with local rural communities and the dubious preservation of biodiversity (Castree 2003).

On the other hand, bioprospecting supporters remark that, in the specific matter of health, bioprospecting can make available new plant species with active principles yet unknown.

2.5 Screening

When a new species is identified as potentially useful for health, two main approaches are practicable, one aiming to the search of a particular biological activity that already exist in the market, the other for the random screening of any biological activity. The first approach tests samples by standard methods; the second demands the realization of a set of bioassays.

The process (Fig. 2.1) starts as soon as the plant species is collected. First, the species has to be identified by a taxonomist, who has to deposit a sample in a botanical repository or herbarium. Immediately after, the plant material is dried, at room temperature or in an oven, preferably avoiding light exposure. Then, it is ground to very small particles or to powder to carry out the extraction of the active principles.

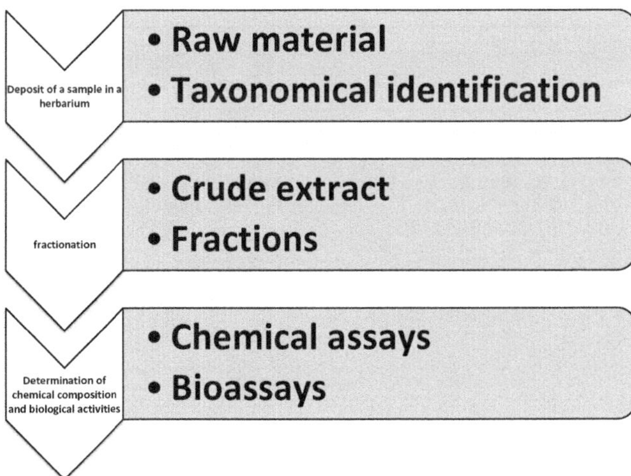

Fig. 2.1 General scheme for the extraction, isolation and characterization of active compounds from plants

2.5.1 *Extractive Process*

When there is no clue about the chemical composition of the species, a preliminary extraction in different solvent systems is performed. An initial step of fractionation is followed by the identification in each fraction of its chemical compounds and activity.

2.5.1.1 Solvent Fractionation

Solvent fractionation is the process of separating the chemical compounds of a crude extract according to their different solubility, basically hydrophobicity/hydrophilicity. In general, extraction is tested in a sequence of solvents of increasing polarity (Katiyar et al. 2012) (Fig. 2.2). The extraction can be performed at room temperature, at high temperatures under reflux, using supercritical fluid or Soxhlet. The extract is then concentrated several times and stored for further studies, in general at low temperatures (–20 °C) to minimize degradation.

2.5.1.2 Chromatographic Fractionation

The chromatographic techniques most used are Thin Layer Chromatography (TLC), High Performance Liquid Chromatography (HPLC), Gas Chromatography (GC),

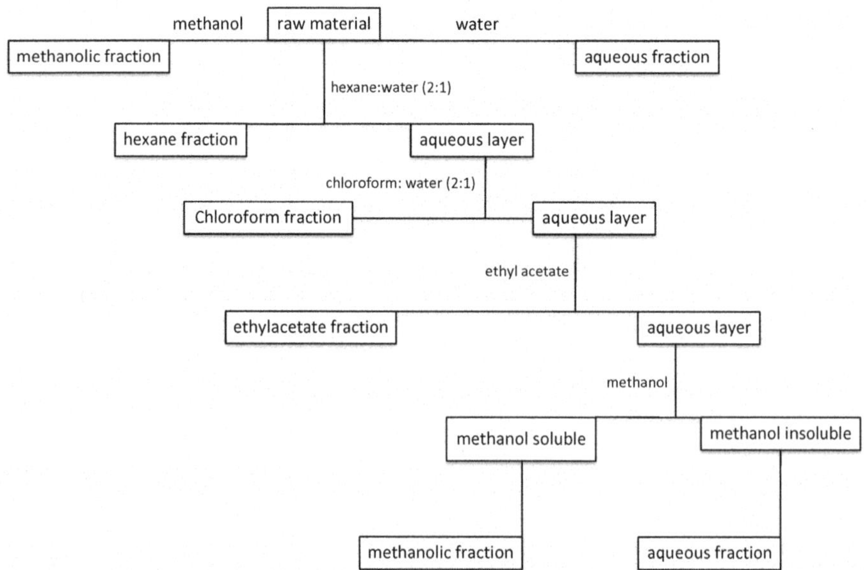

Fig. 2.2 Example of a solvent fractionation scheme

Size Exclusion Chromatography (SEC), etc. In occasions a combination of different techniques and methods are utilized, as HPLC coupled to electron impact-mass spectrometry (Weber et al. 2007).

2.5.1.3 Bioactivity Guided Fractionation

The use of bioassays as fractionation monitors has been the most common strategy to identify bioactive molecules. Limitations to this approach are the presence of interferences, loss of activity during the process of fractionation, and low solubility and instability of the principle of interest (Houghton et al. 2007).

2.5.2 Chemical Screening

Chemical species are identified by standard techniques (Table 2.3). The results provide preliminary information that give information and orientation for the subsequent bioassays. The screening can be performed in the crude plant extracts or in the fractions resulting from the partition through different methods (Tiwari et al. 2011).

2.5.3 Bioassays

The identification of the compounds with biological activity must be performed using sensitive bioassays, which must be simple, reproducible, fast and the least expensive possible. Those assays are performed on the crude extract or in the fractionated extract. Sometimes bioassays are not predictive for clinical efficiency, which makes extremely helpful to perform complementary analysis (Hostettmann 1999).

Table 2.3 Some usual techniques for detecting chemical groups

Chemical group	Standard test
Alkaloids	Mayer's, Wagner's, Draggendorff's, Hager's
Carbohydrates	Molisch's, Benedict's, Fehling's
Glycosides	Modified Borntrager's
Saponins	Froth's, Foam test
Phytosterols	Salkowski's, Liberman Buchard's
Phenols	Ferric chloride test
Tannins	Gelatine test
Flavonoids	Alkaline reagent test, Lead acetate test
Proteins and aminoacids	Xanthoproteic test, Ninhydrin test, Bradford's
Diterpenes	Copper acetate test

Table 2.4 Examples of biological activities tested in plant extracts

Activity	Method
Radical scavengers and antioxidant	Ferric reducing ability of plasma assay (FRAP). Benzie and Strain (1996) Spraying of TLC plates with 2,2-diphenyl-1- picryl hydrazyl (DPPH) radical
Antibacterial	Measurement of inhibition zones on microbial lawns grown in nutrient agar
Antifungal	Microbroth dilution assay following the Guidelines of the National Committee for Clinical and Laboratory Standards (NCCLS) for yeasts M-27-A2 (NCCLS 2002a) and filamentous fungi M-38-A (NCCLS 2002b)
Toxicity	Assay with the crustacean *Artemia salina* Leach. Mortality is tested after 12, 14, 48, 72, 96 and 120 hs. exposition to the plant extract

Plant extracts are tested for different biological activities, e.g.: antioxidant, cytotoxic, bactericidal, antimicrobial, insecticidal, etc. (Ciccia et al. 2000; Mongelli et al. 1997; Desmarchelier et al. 1996), etc. (Table 2.4).

Mclaughlin and Rogers (1998) have adopted an interesting approach with four basic assays for starting the analysis of the biological activity of plant extracts. Those assays are: Brine-shrimp (*Artemia salina*) lethality to test *in vivo* lethality in a simple organism, Crown-gall tumors on potato discs to test antitumor compounds, Frond-inhibition of *Lemna* (duckweed) for testing plant growth stimulants and inhibitors, and the Yellow fever mosquito (YFM) test using eggs of *Aedes aegypti* (Linnaeus) for detecting pesticide activity. Other practical protocols can be found in Liu et al. (2013).

Bioassays can be performed *in vitro*, *in vivo* and *ex vivo*. In all cases it is essential to have an appropriate pharmacological model to establish the applicability of any new drug.

The "omics" technologies can also been helpful to predict the pharmacological and toxic profiles and could replace some of the test regularly performed (Ulrich-Merzenich et al. 2007; Buriani et al. 2012; El-Mowafy 2012).

2.5.3.1 *In Vitro* Assays

They are performed in cells, tissues, microorganisms, insects, mollusks, etc. (Piersen et al. 2004). In general, the approach can be a random screening or a screening based on the traditional knowledge (Agarwal et al. 2014). There is controversy about the extrapolation of *in vitro* results with *in vivo* condition. Several reports have focused on predicting the correlation between both situations (Barnard and Gurevich 2005). In the last years it was suggested as more appropriate a methodology based on system biology or in reverse pharmacology.

2.5.3.2 *In Vivo* Assays

In vivo assays are accomplished employing laboratory animals as guinea pigs, mice, rabbits, etc. Their use must be limited and respecting the principle of the three "R", Replacement, Reduction, and Refinement, of human-animal experimentation and also considering the legislation of each country (Anonymous 2000; Taylor 2010).

2.5.3.3 *Ex Vivo* Assays

Ex vivo experimentation is performed in tissues or cells under environmental conditions with minimal alterations respect to the natural environment of the organism of origin. The experimental conditions are more stringently controlled than *in vivo*.

Ex vivo assays are used for testing biological activities such as anticancer activity, estrogenic activity, and anti-clotting activity (e.g.: ecarin clotting time).

2.6 New Technologies

Some of the difficulties related to the discovering of new molecules for treating human diseases are the lengthy and arduous assays required and consequently their implicit costs. New technologies have been developed to facilitate the process. *In silico* predictions gathers data about drug-target interactions available in public databases (Drug-Bank, KEGG Drug, etc.). Connecting that information with human disease networks can contribute with the finding of new drugs for specific pathologies (Nejad et al. 2013; Spiro et al. 2008). Also, genomics, proteomics, transcriptomics, and metabolomics (the omics) contribute to broaden the knowledge about drug-target interactions. Systems Pharmacology profit of those data to elucidate the therapeutic and adverse effects of drugs (Liu et al. 2013; Nejad et al. 2013).

Polypharmacology integrates the existing information about drugs, targets and diseases to design drugs or cocktails of drugs to treat specific disease states with the lowest toxicity. It requires a multidisciplinary approach among computational modeling, synthetic chemistry, *in vitro*/*in vivo* pharmacological testing, and clinical studies (Reddy and Zhang 2013).

2.7 Concluding Remarks

In the coming years, with the assistance of bioprospecting and ethnobotany, it will be possible to find new medicinal plants still unknown. The holistic approach with *in vivo* and *in vitro* experiments based in chemical and bioassay fractionation will continue contributing to the detection of new compounds (Verpoorte et al. 2006). However, in the last years it became evident that the biological activity of a plant extract is, in general, a consequence of the interaction of the active principles of the plant with other plant chemical components, which is not merely the addition of the activities of each plant chemical but a synergistic effect (Lila and Raskin 2005). A multidisciplinary approach with the contribution of the new technologies will gradually have more impact in the identification of plant bioactive principles and associations for their capacity to find out connections between the activities, mode of action, and toxicity of plant extracts.

References

Agarwal A, D'Souza P, Johnsons TS, Dethe S, Chandrasekaran CV (2014) Use of *in vitro* bioassays for assessing botanicals. Curr Opin Biotechnol 25:39–44

Albuquerque UP (1997) Etnobotanica: uma aproximação teórica e epistemológica. Rev Bras Farm 78(3):60–64

Anonymous (2000) The three Rs declaration of Bologna. Altern Lab Anim 28:1–5

Arenas P (1997) Las fuentes actuales y del pasado para la Etnobotánica del Gran Chaco. Monograf Jard Bot Córdoba 5:17–25

Arenas P (ed) (2012) Etnobotánica en zonas áridas y semiáridas del cono sur de Sudamérica. CEFYBO-CONICET UBA, Buenos Aires

Balunas MJ, Kinghorn AD (2005) Drug discovery from medicinal plants. Life Sci 78:431–441

Barnard R, Gurevich KG (2005) *In vitro* bioassays as predictors of *in vivo* response. Theor Biol Med Model 2:3, http://www.tbiomed.com/content/2/1/3

Benzie IF, Strain JJ (1996) The ferric reducing ability of plasma (FRAP) as measurement of "antioxidant power": the frap assay. Anal Biochem 239:70–76

Buriani A, Garcia-Bermejo ML, Bosisio E, Xu Q, Li H et al (2012) Omic techniques in systems biology approaches to traditional Chinese medicine research: present and future. J Ethnopharmacol 140:535–544

Castree N (2003) Bioprospecting: from theory to practice (and back again). Trans Inst Br Geogr 28:35–55

Ciccia G, Coussio J, Mongelli E (2000) Insecticidal activity against *Aedes aegypti* larvae of some medicinal South American plants. J Ethnopharmacol 72:185–189

Desmarchelier C, Mongelli E, Coussio J, Ciccia G (1996) Studies on the cytotoxicity, antimicrobial and DNA-binding activities of plants used by the Ese'ejas. J Ethnopharmacol 50:91–96

Doughari JH (2012) Phytochemicals: extraction methods, basic structures and mode of action of potential chemotherapeutic agents. In: Rao V (ed) Phytochemicals – a global perspective of their role in nutrition and health. InTech Croatia. Open Access, pp 1–32

El-Mowafy AM (2012) Herbal therapy: can omics technology create order from chaos? Biochem Anal Biochem 1:e130

Gurib-Fakim A (2006) Medicinal plants: traditions of yesterday and drugs of tomorrow. Mol Aspects Med 27:1–93

Harvey AL, Gericke N (2011) Bioprospecting: creating a value for biodiversity. In: Pavlinov I (ed). Research in Biodiversity - Models and Applications InTech Croatia. Available from: http://www.intechopen.com/books/research-in-biodiversity-models-and-applications/bioprospecting-creating-a-value-for-biodiversity

Hostettmann K (1999) IUPAC, Invited lecture presented at the international conference on biodiversity and bioresources: conservation and utilization, 23–27 Nov 1997, Phuket, Thailand. Strategy for the biological and chemical evaluation of plant extracts. http://www.iupac.org/symposia/proceedings/phuket97/hostettmann.html

Houghton PJ, Howes MJ, Lee CC, Steventon G (2007) Uses and abuses of *in vitro* tests in ethnopharmacology: visualizing an elephant. J Ethnopharmacol 110:391–400
Jones V (1941) The nature and scope of Ethnobotany. Chron Bot 6(10):219–221
Katiyar G, Gupta A, Kanjilal S, Katiyar S (2012) Drug discovery from plant sources: an integrated approach. Ayu 33(1):10–19
Lila MA, Raskin I (2005) Health-related interactions of phytochemicals. J Food Sci 70:R20–R27
Liu J, Pei M, Zheng C, Li Y, Wang Y, Lu A, Yang L (2013) A system-pharmacology analysis of herbal medicines used in health improvement treatment: predicting potential new drugs and targets. Evid Based Complement Altern Med 2013, Article ID 938764, 17 pp. http://dx.doi.org/10.1155/2013/938764
Mclaughlin J, Rogers L (1998) The use of biological assays to evaluate botanicals. Drug Inf J 32:513–524
Molares S, González S, Ladio A, Castro MA (2009) Etnobotánica, anatomía y caracterización físico-química del aceite esencial de *Baccharis obovata* Hook. et Arn. (Asteraceae: Astereae). Acta Bot Bras 23(2):578–589
Mongelli E, Desmarchelier C, Coussio J, Ciccia G (1997) Biological studies of *Bolax gummifera*, a plant of the Falkland Islands used as a treatment of wounds. J Ethnopharmacol 56:117–121
Moronkola DO (2012) Developments in phytochemistry, drug discovery research in pharmacognosy. In: Vallisuta O (ed). Drug Discovery Research in Pharmacognosy In Tech Croatia. Available from: http://www.intechopen.com/books/drug-discovery-research-in-pharmacognosy/developments-inphytochemistry
NCCLS (2002a) Application of a quality management system model for laboratory services: approved guideline GP26-A3. NCCLS, Wayne
NCCLS (2002b) Clinical technical procedures manual. Approved guideline GP2-A2. NCCL, Wayne
Nejad AM, Mousavian Z, Bozorghmer JH (2013) Drug-target and disease networks: polypharmacology in the post-genomic era. In Silico Pharmacol 1:17, http://www.in-silico-pharmacology.com/content/1/1/17
Payne GF, Bringi V, Prince C, Shuler M (1991) The quest for commercial production of chemicals from plant cell culture. In: Payne GF, Bringi V, Prince C, Shuler ML (eds) Plant cell and tissue culture in liquid systems. Hanser, Munich, pp 1–10
Piersen CE, Booth NL, Sun Y, Liang W, Burdette JE et al (2004) Chemical and biological characterization and clinical evaluation of botanical dietary supplements: a phase I red clover extract as a model. Curr Med Chem 11:1361–1374
Pirondo A, Keller HA (2012) Raúl N. Martínez Crovetto: los albores de la etnobotánica en Argentina Introducción a estudios etnobotánicos V. Bonplandia 21(2):101–107
Raskin I, Ribnicky DM, Komarnstsk S, Ilic N, Puolev A et al (2002) Plants and human health in the twenty-first century. Trends Biotechnol 20(12):522–531
Reddy AS, Zhang S (2013) Polypharmacology: drug discovery for the future. Expert Rev Clin Pharmacol 6(1). doi:10.1586/ecp.12.74
Rosso CN (2013) La Etnobotánica histórica: el caso Mocoví en la reducción de San Javier en el siglo XVIII. Etnobiologia 11(3):54–65
Scarpa G, Montani M (2011) Etnobotánica médica de las "ligas" (*Loranthaceae* sensu lato) entre indígenas y criollos de Argentina. Dominguezia 27(2):5–19
Schmidt B, Ribnicky D, Pulev A, Logendra S, Cefalu W, Raskin I (2008) A natural history of botanical therapeutics. Metab Exp 57(Suppl Clinical and 1):S3–S9
Spiro Z, Kovacs IA, Csermely P (2008) Drug-therapy networks and the predictions of novel drug targets. J Biol 7:20
Suárez ME (2009) El análisis de narrativas en etnobotánica: el "yuchán" (Ceiba Chodatii, Bombacaceae) en el discurso de los wichís del Chaco Semiárido salteño, Argentina. Bol Soc Argent Bot 44(3–4):405–419
Taylor K (2010) Reporting the implementation of the Three Rs in European primate and mouse research papers: are we making progress? Altern Lab Anim 38:495–517

Teiten MH, Gaascht F, Dicato M, Diedrich M (2013) Anticancer bioactivity of compounds from medicinal plants used in European medieval traditions. Biochem Pharmacol 86:1239–1247

Tiwari P, Kumar B, Kaur M, Kaur G, Kaur H (2011) Phytochemical screening and extraction: a review. Internationale Pharmaceutica Sciencia 1(1):98–106

Ulrich-Merzenich G, Zeitler H, Jobst D, Panek D, Vetter H, Wagner H (2007) Application of the "-Omics-" technologies in phytomedicine. Phytomedicine 14:70–82

UN (1992) Convention on the biological diversity. Rio de Janeiro, http://www.biodiv.org/doc/legal/cbd-en.pdf

Verpoorte R, Kim H, Choi Y (2006) Plants as source of medicines. New perspectives. In: Bogers RJ, Craker LE, Lange D (eds) Medicinal and aromatic plants. Springer, Dordrecht, pp 261–273. Printed in the Netherlands

Vines G (2004) Herbal harvests with a future: towards sustainable sources for medicinal plants. Plantlife International. www.plantlife.org.uk

Weber M, Knoy C, Kindscher K, Brown RCD, Niemann S, Chapman J (2007) Purification of medicinally active compounds in Prairie plants by HPLC coupled to Electron Impact-Mass Spectrometry. Am Lab 39(12):9–11

Zardini A (1984) Etnobotánica de Compuestas Argentinas con Especial Referencia a su Uso Farmacológico. Acta Farm Bonaerense 3(1):77–99

Chapter 3
Plant Secondary Metabolism

Abstract Plant metabolism is divided in primary and secondary metabolism according to the functions they display and the characteristics of their biosynthetic pathways. Secondary metabolites are synthesized from primary metabolites and, usually, they are produced within a phylogenetic group, but are not only limited to them.

In general, secondary metabolites display a role in the adaptation of plants to their environment, e.g.: in defense against predators and pests. They are diverse chemical structures broadly classified in three groups: phenolics, terpenes and steroids, and alkaloids. The two first groups have in common an anti-oxidant and anti-inflammatory activity besides the specific biological activities typical of each group. Alkaloids are a heterogeneous group with a variety of actions on the central and peripheral nervous system, among others.

Metabolic engineering has allowed deciphering and modulating some of the metabolic pathways of these metabolites. Also, it is an interesting tool that helps to improve secondary metabolite production of commercial interest. New approaches as systems biology and synthetic biology will help to discover interactions between secondary metabolites and human pathologies and to obtain better secondary metabolite productivities.

Keywords Plant secondary metabolism • Secondary metabolites • Metabolic engineering • Synthetic biology

3.1 Introduction

Plant chemical compounds can be classified as primary or secondary metabolites according to their biosynthetic pathway and functions. The biosynthesis of primary metabolites practically does not differ among the living organisms, playing an essential role for life (growth, metabolism, and reproduction). On the other hand, secondary metabolites are organic molecules not essential for plant growth and development but with important roles in the relationship with the environment as in plant defence and the adaptation of plants to their surroundings (Bernohf 2010; Bourgaud et al. 2001; Harborne 1999).

M.A. Alvarez, *Plant Biotechnology for Health: From Secondary Metabolites to Molecular Farming*, DOI 10.1007/978-3-319-05771-2_3

3.2 Primary and Secondary Metabolism

Plants, and some bacteria, are able to use the solar energy, plus carbon dioxide and water, to synthesize complex molecules (carbohydrates as sucrose and starch) through photosynthesis, a process that generate chemical energy (ATP) to be used in cellular metabolism. Anaerobic and aerobic organisms make use of that chemical energy, with different efficiency, in their specific glycolytic process. In plants, the glucose derived from those carbohydrates produce phosphoenol pyruvate (via glycolysis) and malate (via pentose phosphate shunt).

Secondary metabolites are synthesized from primary metabolites by a sequence of chemical reactions catalyzed by enzymes some of which limits the pathway defining the synthesis and the amount of metabolite produced.

Malate enters into the shikimate pathway to produce tryptophan, from which originate indole alkaloids, and phenylalanine.

Glycolysis take place in the cytoplasm of the plant cell, the obtained phosphoenol pyruvate enters into the mitochondria and in its stroma produces pyruvate, which is transformed into aceto acetyl-CoA. Aceto acetyl-CoA goes into the cycle of the tricarboxylic acid (TCA) in the mitochondrial crests, completing the catabolism of glucose through a series of oxidative reactions, with the production of carbon dioxide and adenosine triphosphate (ATP).

Several molecules, intermediate in the synthesis of secondary metabolites, are produced in the different steps of the TCA, proline that originates pyrrolic alkaloids, fatty acids from which derive phospholipids, triacylglycerol, glycosides, phenols, mevalonate, ornitine, and cinnamic acid. Mevalonate, by a series of intermediate chemical reactions, originates carotenoids, stigmasterol and terpenoids. Orinitine together with phenylalanine (coming from the shikimate pathway) are the precursors of tropane alkaloids, and cinnamic acid originates flavonoids and lignans (Fig. 3.1).

3.3 Secondary Metabolites

Secondary metabolites are diverse chemical structures broadly classified in three groups: phenolics, terpenes and steroids, and alkaloids (Bourgaud et al. 2001; Harborne 1999) (Fig. 3.2). A complex network of transcription factors controls their transcriptional response to developmental or environmental stimuli (Patra et al. 2013; Petinatti Pavarini et al. 2012; Vom Endt et al. 2002). Most of them have an ecological role in regulating the interactions between plants, microorganisms, insects and animals. They often accumulate in the plant in small quantities, frequently in specialized cells, in specific plant developmental steps, or during stress (Piñol and Palazón 1993).

A group of secondary metabolites are high-molecular-weight polymeric compounds such as cellulose, lignanes, etc., which are distributed in all higher plants having mainly a structural function. Others are low-molecular-weight compounds with a more constrained distribution, in general synthesized in response to the exposition to some biotic or abiotic substances (Dixon 2001). Frequently, they are produced within a phylogenetic group, but they are not only limited to it.

In planta they can be defensive substances, antifeedants, attractants or pheromones. *Ex planta*, they display a variety of biological activities that are exploited by people in different manners (Table 3.1).

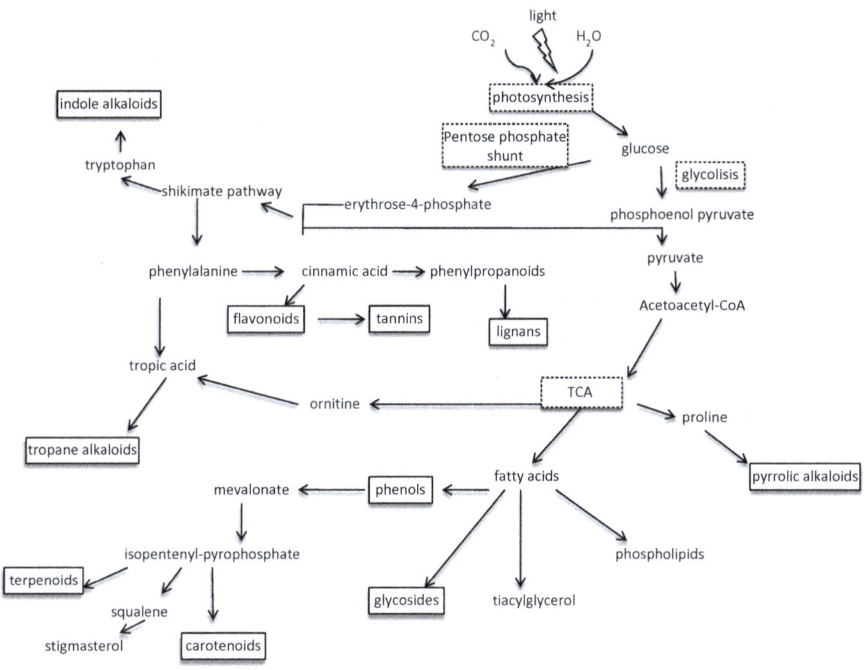

Fig. 3.1 Primary and secondary metabolism in plants

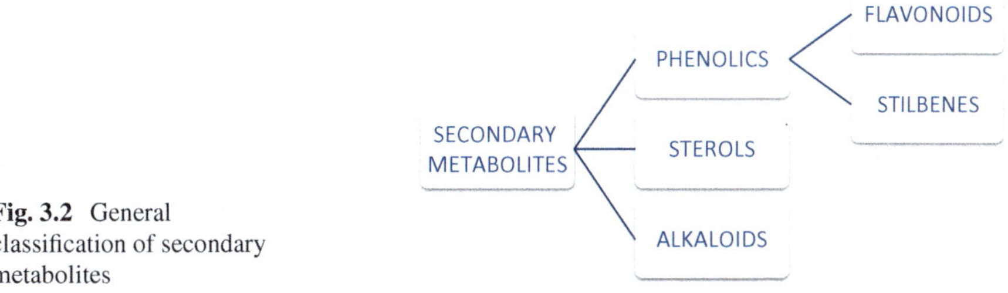

Fig. 3.2 General classification of secondary metabolites

Table 3.1 Plant secondary metabolites with commercial relevance

1. Dyes, flavours and fragrances
Dyes: anthocianins, betacianins, catechu, haematoxylin, gamboge, indigo, rubhada, saffron, madder, weld
Flavours: asparagus, celery, cinnamomum, strawberry, vanilla
Fragrances: cinnamon, capcaicin, eucalyptus, garlic, jasmine, lemon, mint, onion, patchouli, rosewood, sassafras, sandalwood, vetiver. Sweeteners: steviosids, taumatin, miraculin, monellin
2. Agrochemicals
Azadiachtin, dictamine, ecdysterone, harmaline, indanediones, neriifolin, physostigmine, piretrines, polygodial, rotenone, ryanodine, salannin, wyerone
3. Pharmaceuticals
Ajmalicine, artemisinin, atropine, codeine, digoxin, diosgenin, hyoscyamine, L-dopa, morphine, paclitaxel, quinine, scopolamine, serpentine, shikonin, vinblastine, vincristine

3.3.1 Phenolics

Some of the representatives of this group have structural functions (e.g.: lignin reinforcing specialized cell walls). Others participate in plant defense (e.g.: coumarins, condensed tannins, isoflavonoids, proanthocyanidines, stilbenes), pollination, light screening, and seed development.

Chemically, they are characterized as aromatic compounds with one or more acidic hydroxyl groups. Exposed to the air they quickly oxidize and complex with proteins.

Most phenolic compounds derive from the phenylpropanoid and phenylpropanoid-acetate pathway, being phenylalanine ammonia lyase (PAL) a key enzyme in the pathway. Others derive from alternative pathways, e.g.: tetrahydrocannabinoids are polyketide or terpenoid derivatives. Phenolics are classified in Flavonoids and Stilbens.

3.3.1.1 Flavonoids

Flavonoids are a subgroup of polyphenolic compounds with a characteristic C6–C3–C6 structure. The chemical structure of flavonoids is based on a C15 skeleton bearing a second aromatic ring B in position 2, 3 or 4 (Fig. 3.3). Flavonoids accept several substituents (hydroxyl-, methyl- groups, isopentenyl units, sugars) that are determinants of their solubility. They are a heterogeneous family whose main subclasses are anthocyanidins, flavanones, flavan-3-ols, flavones, flavonols, and isoflavones. Minor groups are aurones, coumarins, chalcones, dihydrochalcones, dihydroflavonols, and flavan-3,4-diol.

They are often located in the vacuole of cells, and are responsible of the color of most of flowers and fruits, functioning as pollinator attractants (pelargonidins, cyanidins, delphinidins), protecting plants from UV-B radiation (kaempferol), or

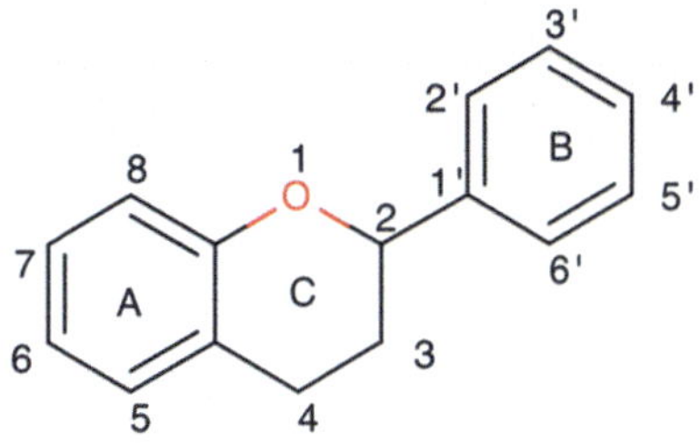

Fig. 3.3 Chemical structure of flavonoids based on a C15 skeleton

acting as insect-feeding attractants (isoquercetine) or antifeedants (proanthocyanidins). Also, some of them (apigenin and luteoline) participate as signal molecules in the legume-*Rhizobium* association.

Many flavonoids are synthesized by the stimuli of stress as wounding, drought, and exposition to metals and starvation.

Flavonoids, especially flavonols and flavan-3-ols, have been suggested to protect plants from oxidative stress acting as antioxidants (Hernández et al. 2009).

As for their biological activities, they display anti-oxidant function, anti-inflammatory effect (Comalada et al. 2006; Fang et al. 2005), and inhibition of cell-signal proteins (Pang et al. 2006). The Malvidin-3-O-β-glucoside, found in grape, inhibits mediators of macrophage-derived inflammation (Decendit et al. 2013). They were also considered as beneficial against certain types of cancer for their activity, demonstrated by *in vitro* and *in vivo* assays, as modulators of apoptosis, vascularization, cell differentiation, cell proliferation, etc. The cross talk between flavonoids and the key enzymes related to neoplastic cells and metastasis has not yet been elucidated (Batra and Sharma 2013).

3.3.1.2 Stilbenes

They are a small group of phenolic metabolites that participate in plant defense. Resveratrol is the most common stilbene (Fig. 3.4); it could be present as the *cis* or the *trans* isomer and are produced in response to stress (wounding or pathogen attack). *Trans*-resveratrol is the most abundant of both isomers in plant tissues and is considered beneficial against coronary disease in humans for its antioxidant activity (Bradamante et al. 2004). Also, it is though that it has a protective activity against certain kind of cancer (Aggarwal et al. 2004; Kundu and Surh 2008), diabetes (Sharma et al. 2007), and neuro-degenerative diseases (Anekonda 2006; Pearson et al. 2008) through its induction of *Sirtuin-1* genes (Donnez et al. 2006). *Trans*-resveratrol is currently produced from *Polygonum cuspidatum* (syn. *Fallopia japonica*) for being used in health care or from grapevine for cosmetics, among other possible sources (Burns et al. 2002; Counet et al. 2006).

Phytoestrogens are flavonoids with structural similarities to mammalian 17 β-estradiol (E2) (e.g.: resveratrol, genistein, kaempferol, daidzein). They have demonstrated their ability of binding to estrogen receptors (Pilsakova et al. 2010) and,

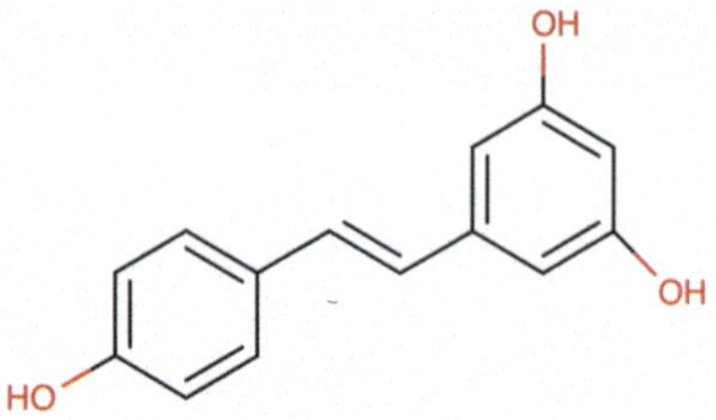

Fig. 3.4 Chemical structure of resveratrol

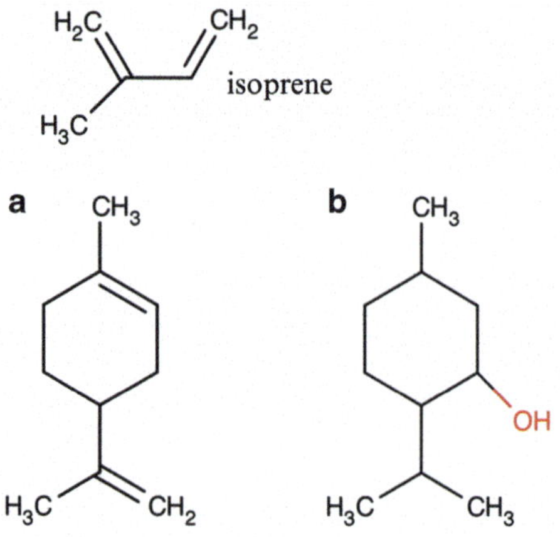

Fig. 3.5 Chemical structure of terpenes (**a**) limonene, (**b**) menthol

in a similar way as E2, are competent to alleviate postmenopausal complaints, increase bone formation and repress adipose tissue with the advantage of not producing the estrogens induced side effects typical of the hormone replacement therapy (Schilling et al. 2014; Song et al. 2007; Heim et al. 2004).

3.3.2 *Terpenes and Steroids*

3.3.2.1 Terpenes

They are a group derived from the repetitive "head to tail" (although "head to head" and "head to middle" unions are also possible) fusion of a branched unit of five carbons, the isoprene (Fig. 3.5). They are classified, according to their number of five-carbon units, as hemiterpenes (C1), monoterpenes (C10), sesquiterpenes (C15), diterpenes (C20), triterpenes (C30), tetraterpenes (C40), and polyterpenes (more than 80-carbon units). Plants produce a wide variety of terpenoids in specialized structures such as glandular trichomes, secretory cavities, and glandular epidermis. In general, the C15, C30, and polyterpenes are produced in the cytosol and endoplasmic reticulum (acetate/mevalonate pathway) whereas the isoprene, C10, C20

Table 3.2 Terpenes of commercial application

Terpene	Source	Use
Geraniol	*Pelargonium graveolens*	Scent
Citral	*Cymbopogon flexuosus*	Scent
Menthol	*Mentha arvensis*	Flavour
Camphor	*Cinnamomum camphora*	Mothproofing treatment
Caryophyllene	*Syzygium aromaticum, Cannabis sativa*	Spice, anti-inflammatory
Eugenol	*Syzygium aromaticum*	Spice
Santonin	*Artemisia maritima*	Anti-parasitic
Artemisinin	*Artemisia annua*	Anti-malarial
Abietic acid	*Pinus* spp., *Abies* spp.	Varnishing, resin soap
Pimaric acid	*Pinus* spp.	Resin
Stevioside	*Stevia rebaudiana*	Sweetener
Taxol	*Taxus brevifolia*	Antitumour activity
Azadirachtin	*Azaridachta indica*	Insect anti-feedant
Diosgenin solasodine	*Dioscorea* spp, *Solanum eleagnifolim.*	Synthesis of steroidal hormones
Digitalis glycosides	*Digitalis lanata*	Cardiotonic

and C40 are produced in plastids (glyceraldehyde phosphate/pyruvate pathway) (Croteau et al. 2000). Most of the terpenes have cyclic structures.

Monoterpenes are major components of the aromas of the plants, the essential oils, which are used in perfumery and flavouring industries (Table 3.2).

3.3.2.2 Steroids

Steroids have a cyclopentane perhydrophenanthrene backbone, and derive from tetracyclic terpenes. Ergosterol and stigmasterol are found in the plant kingdom, while some plant steroids, such as the digitalis glycosides (cardenolides) from *D. lanata* and solasodine from *S. eleagnifolium*, can be used to synthesize animal steroidal hormones. The most abundant phytosterols are β-sitosterol, campesterol and stigmasterol (95–98 % of all plant sterols) (Fig. 3.6). They have cholesterol-lowering properties and have a protective effect against cardiovascular disease (Sabater-Jara and Pedreño 2013; García-Llatas and Rodríguez-Estrada 2011). Also, they have anti-atherosclerotic, anti-inflammatory and anti-oxidative activities (Delgado-Zamarreño et al. 2009), and antitumor effect against certain types of cancer (Woyengo et al. 2009).

Brassinosteroids are natural plant growth regulators that promote cell expansion and elongation, vascular differentiation, reorientation of microtubules, inhibition of root elongation, and senescence (Bishop and Koncz 2002).

Fig. 3.6 Chemical structure of stigmasterol

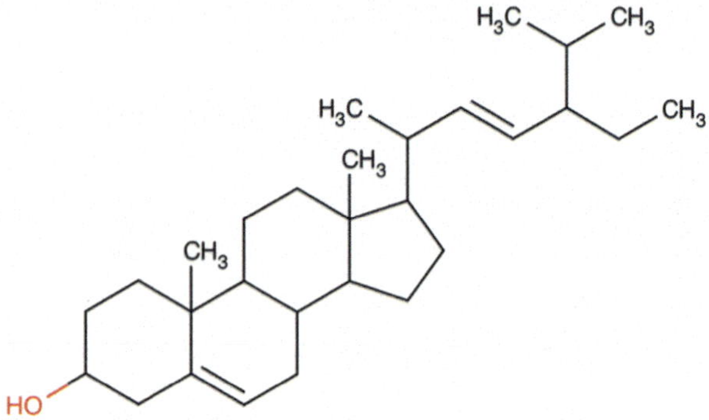

Carotenoids provide colour to plants as β-carotene and lycopene that brings their color to carrot and tomato respectively. They are anti-oxidants and precursors in the synthesis of vitamin A.

Saponins are steroidal or triterpenic glycosides some of which were reported as having similar activities as phytoestrogens (e.g.: PNS from *Panax notoginseng*) (Augustin et al. 2011). It was reported that, as phytoestrogens, some saponins are able to influence the balance between osteogenic and adipogenic differentiation of mesenchymal stem cells. In doing so, they would favor or even stimulate osteogenic differentiation, and as a consequence would aid to prevent bone loss and fragility fractures (Schilling et al. 2014). Other saponins, as tigogenin and platycodin, were associated to the prevention of obesity, inhibition of adipogenesis, and promotion of osteogenesis; consequently they are proposed as potentially useful as an alternative treatment in disorders as osteoporosis (Lee et al. 2010; Zhou et al. 2007).

3.3.3 Alkaloids

People have been using alkaloids since centuries, in magical or curative potions, as painkillers, poisons, etc. The use of opium, the latex of *Papaver somniferum* containing codeine and morphine, is already reported in de Middle East as far as 1200–1400 B.C.

Initially, alkaloids were chemically described as basic molecules that contain one or more nitrogen atoms produced by plants (Fig. 3.7). Nowadays it is known that some alkaloids are also produced by certain animals such as the European fire salamander that produces samandarines, frogs and toads that probably overexpress an uptake system for sequestration of dietary arthropod alkaloids into skin glands, or Australian myobatrachid frogs of the genus *Pseudophryne* that synthesize their own unique indolic pseudophrynamines while sequestering pumiliotoxins (PTXs) from a dietary source (Daly et al. 2002).

Alkaloids are molecules capable to form salts with acids and to complex with metal ions. They can be classified based on different characteristics, e.g.: according

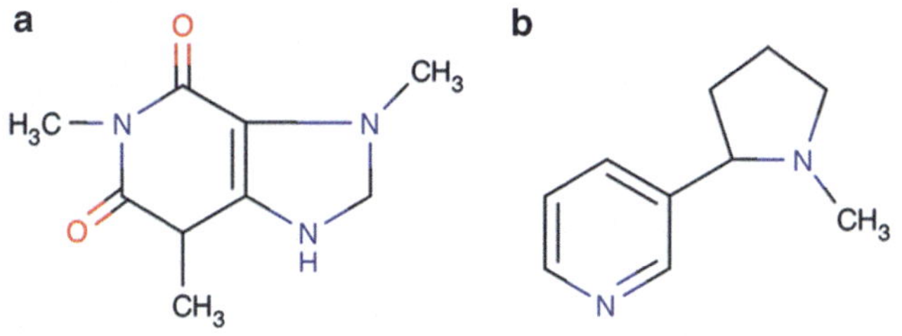

Fig. 3.7 Chemical structures of (**a**) nicotine and (**b**) caffeine

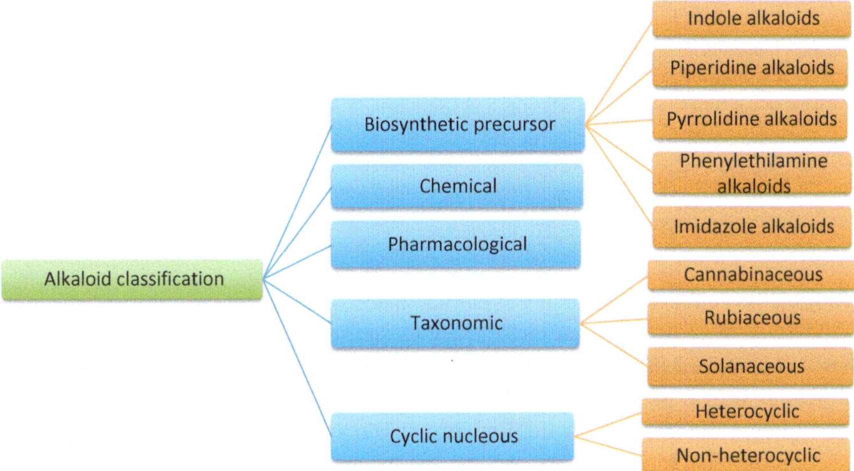

Fig. 3.8 Alkaloid classification according different characteristic and properties

to the aminoacids from which derive, according the structure of the ring system containing the nitrogen atom (pyrrolidine-, indole-, piperidine-, pyrrolizidine-, tropane-, quinoline-, isoquinoline-, aporphine-, imidazole-, diazocin-, purine-, steroidal-, amino-, and diterpene alkaloids), according their pharmacological activities (analgesic, narcotic, etc.), according the taxonomic classification of the plant family where they are mostly found (Cannabinaceous, Rubiaceous, Solanaceous), or the characteristics of the cyclic nucleus (heterocyclic, non-heterocyclic) (Fig. 3.8). About 20 % of plants produce alkaloids; many of them produce more than one type.

The role of alkaloids in plants is mostly in defense against predators as insects (nicotine, caffeine) and herbivores (lupine alkaloids, solasodine), and against microorganisms (berberine, liriodenine) as well. Also, some of them are synthesized in response to tissue damage (nicotine). Quite a lot of them have biological activity (Table 3.3).

Most alkaloids derive from amino-acid precursors (i.e. ornithine, lysine, tyrosine, tryptophan, and histidine) or from anthranilic acid or nicotinic acid. Alkaloid bio-

Table 3.3 Plant alkaloids with biological activity

Alkaloid	Source	Biological activity
Arecoline	*Areca catechu*	Stimulant
Atropine	*Atropa belladonna*	Anticholinergic
Berberine	*Berberis spp.*	Pigment, mild antibiotic
Caffeine	*Coffea arabiga*	Stimulant
Cocaine	*Erythroxylon coca*	Anesthesic, stimulant
Coniine	*Conium maculatum*	Paralysis of nerve endings
Ephedrine	*Ephedra sinica*	Bronchodilator
Mescaline	*Lophophora williamsii*	Hallucinogenic
Nicotine	*Nicotiana tabacum*	Neuroactive, insecticide
Papaverine	*Papaver somniferum*	Muscle relaxant
Piperine	*Piper nigrum*	Spice
Quinine	*Cinchona ledgeriana*	Antimalarial, bitter
Reserpine	*Rauwolfia*	Antipsychotic
Strichnine	*Strychnos*	Poison
Theophylline	*Camelia sinensis*	Stimulant
Vinblastine	*Catharanthus roseus*	Antineoplasic

synthesis is regulated in time by development and environmental factors controlling the activity of transcriptional factors, which program the whole process. Usually, they are produced in specialized cells, tissues, and organs, where they accumulate or they are transported from the place of synthesis to the place of storage (De Luca and St-Pierre 2000; De Luca and Laflamme 2001).

Catharanthus roseus has been taken as a model system, to elucidate that the biosynthesis of the important anticancer agents vinblastine and vincristine can be regulated by a number of biotic and abiotic stimuli being activated at particular stages of plant development (Shanks and Morgan 1999; Facchini 2001).

3.4 Metabolic Engineering

It was originally defined as 'the direct improvement of production, formation, or cellular properties through the modification of specific biochemical reactions or the introduction of new ones with the use of recombinant DNA technology' (Bailey 1991).

Plant metabolic engineering plays a special role in biotechnological approaches for covering the escalating demands of food, energy and medicines of an ever-growing world population (Dudareva and DellaPenna 2013).

Genetic engineering of secondary metabolite biosynthesis is intended to increase the amount of a compound of interest or decrease the amount an unwanted metabolite in plants. To increase the amount of a specific metabolite, or to incorporate the synthesis of novel compounds, the inhibition of competitive pathways, the reduction of the catabolism of the target compound, and the modification of the expression (over expression or suppression) of regulatory genes are approaches that have been explored. On the other hand, to reduce the amount of an undesirable compound the approach can be the knock out of a certain enzymatic step by RNA antisense, RNA co-suppression or RNA interference, or the over expression of an antibody against any key enzyme in the pathway. Also, diverting the pathway into a competitive one, or increasing the catabolism of the undesired product has been used (Verpoorte and Memelick 2002). Even though plant metabolic engineering has had some success the modifications of chemical profiles is complex and mostly an ineffective strategy (Jirschitzka et al. 2013).

Up to date, the best-studied pathway at genetic level is that from flavonoids and anthocyanins. Also, there are great advances in the elucidation of the indole and isoquinoline alkaloid pathways (Allen et al. 2004; Runguphan and O'Connor 2009; Glenn et al. 2011; Park et al. 2003).

In plants, transcription factors (TF) regulate the expression of multiple genes in a single biosynthetic pathway. TFs from the biosynthesis of defense compounds are mainly from the bZIP, MYB, MYC, WRKY and ERF families. In the case of anthocyanin and tannin pathways, MYB family transcription factors are fundamental for the coordinated activation of genes (Dixon et al. 2013). Jirschitzka et al. (2013) summarize the TFs that were lately employed in the engineering of plant defence pathways and that were used to manipulate the biosynthesis of e.g.: proanthocyanidin in poplar, artemisinin in *A. annua*, nicotiana in *N. tabacum*, sesquiterpenes in *A. thaliana* (Hong et al. 2012), terpenoid indole alkaloid in *C. roseus* (Suttipanta et al. 2011). What is more, TFs from a species have demonstrated to be effective in regulating the biosynthetic pathway in other species; e.g. AtMYB12 from *A. thaliana* also increases the concentration of phenylpropanoids and flavonoids in *N. tabacum* (Misra et al. 2010).

A limitation on the use of transcription factors is their possible effects on multiple pathways that could lead to unwanted phenotypes.

Multigene engineering (Christou 2013) has also helped to the development of crops with more complex traits, including extended metabolic pathways producing valuable compounds such as β-carotene in Golden Rice (Ye et al. 2000) and three different vitamins (β-carotene, ascorbate, and folate) representing three different metabolic pathways in Multivitamin Corn (Naqvi et al. 2009).

When engineering a metabolic pathway, promoters have also to be considered. Certain compounds are only produced in specific organs, tissues, or specific storage structures because it is the site of accumulation of their precursors, because in that place they are less susceptible to degradation, or plant tissues are protected from their toxicity. Selecting the adequated promoter is thus fundamental for the expression of the involved genes. In the case of multigene constructs, genes can be under the same promoter (Moldrup et al. 2012; Geu-Flores et al. 2009a) or each gene can have a different promoter (Farhi et al 2011). Also, constitutive (Samac et al. 2004) or organ-specific promoters can be used (Redillas et al. 2012; Bai et al. 2011), being the last strategy more adequate for avoiding spurious side products, to improve yield, avoid autotoxicity, and increase crop protection.

A typical example is the biosynthesis of monoterpene indole alkaloids in *C. roseus* (Heinig et al. 2013) where the metabolism takes place across various cellular locations, including the chloroplast, vacuole, nucleus, endoplasmic reticulum, and cytosol (Verma et al. 2012). Thus, when engineering a particular biosynthetic pathway it is important to take into account this information because the ignorance of which specific enzyme participates at a defined localization can be determinant of the failure.

Data mining is today a prerequisite to a successful strategy (Giddings et al. 2011; Hagel and Facchini 2010; Winzer et al. 2012; Liscombe et al. 2010). Bioinformatics play a remarkable role in offering all the information needed about co-expression analyses, comparative metabolite profiling, transcription factors and enzymes at metabolic bottlenecks.

To perform gene-stacking, RNA-based silencing systems, including RNAi (Runguphan et al. 2009), the use of synthetic RNA elements, ribosome binding site elements, and a combination of promoters strategically placed in front of stacked pathway genes could hypothetically facilitate to modulate protein expression (Hawkins and Smolke 2008; Chang et al. 2012; Glenn et al. 2013).

The intrusion in the secondary metabolite to favor the biosynthesis of a particular compound has to take into account not to deplete the primary metabolism.

3.5 Synthetic Biology

The main goal of synthetic biology is to modify a biosynthetic pathway in an organism in order to increase or produce a specific compound (Kliebenstein 2014). The establishment of multigene/protein-based networks has enabled the construction of novel synthetic pathways (Purnick and Weiss 2009). The integration of multiple genes involved in a metabolic pathway into a single vector under different promoters was already used for producing carotenoids, vitamin E and PUFAs (Naqvi et al. 2010), glucosinolates (Geu-Flores et al. 2009b) and artemisinin (Farhi et al. 2011).

The fact that some secondary metabolites are stored in different sites than those of biosynthesis necessitates intracellular and intercellular transport. However, there are only very few reports about transport engineering in synthetic biology (Goossens et al. 2003), possibly due to the limited molecular knowledge on intracellular transport and cellular excretion of specialized metabolites (Nour-Eldin and Halkier 2013). Glucosinolates are an excellent model system for studying transport of specialized metabolites (Nour-Eldin and Halkier 2009).

3.6 Systems Biology

Systems biology is about the data integration and the use of the omics technologies to understand the cell at a system level (Noble 2008; Sagt 2013). Also, it considers a biological system feasible to be modeled as a network (Bebek et al. 2012). Network analysis can be classified in: omics data modeling (stoichiometric modelling, kinetic modeling), which is the use of statistical methods to identify and infer complex functional interactions among the components in biological systems (Chae et al. 2012), stoichiometric modelling which predicts flux distributions of biological pathways (Schellenberger et al. 2011), and kinetic modeling, which is utilized for the evaluation of the dynamics of biological systems such as time-course simulation, steady-state analysis and metabolic control analysis (Rohwer 2012). From them, omics data modeling is the most commonly used for predicting gene function and comprehend a whole biological system, allowing the elucidation of various networks as metabolite-to-metabolite, gene-to-metabolite and protein-protein interactions (Vidal et al. 2011) (Yonekura-Sakakibara et al. 2013). The combination of plant metabolomics, plant systems biology, and bioinformatics has generated and assembled enough data to model complex plant processes (Christou 2013).

The methods of Network flux analysis (NFA) differ from one another in the size and complexity of the network under analysis and in other factors as the type of information, insight and predictions they make (Shachar-Hill 2013). Plant metabolic networks are more complicated and numerous, with the additional factor that they have several metabolic flux options in a pathway or among the different cells of a specific organ in a determined step of plant development. A good example of NFA application in plant metabolism is the dynamic modeling of monoterpene biosynthesis in peppermint. The hypothesis of an undiscovered regulatory inhibition of a key enzyme was tested *in silico* and *in vitro* finding a good agreement between the simulated and experimental monoterpene accumulation profiles and feedback inhibition (Rios-Estepa et al. 2008).

3.7 Conclusive Remarks

The plant kingdom has a great proportion of species yet unknown or whose chemical composition has not been described. Secondary metabolites display a diverse set of biological activities of therapeutic use; the finding of new plant species or plant compounds could increase that battery of chemicals for fighting diseases. The production at large-scale of secondary metabolites can be improved by optimization the process of production, harvest and down-stream processing of metabolites from crops or by using the technology of production in *in vitro* cultures (phytofermentations). Combining those strategies with the new approaches of systems and synthetic biology will help to discover interactions between drugs and pathologies and to obtain better productivities.

References

Aggarwal BB, Bhardwaj A, Aggarwal RS, Seeram NP, Shishodia S, Takada Y (2004) Role of resveratrol in prevention and therapy of cancer: preclinical and clinical studies. Anticancer Res 24:2783–2840

Allen RS, Milligate AG, Chilty JA, Thisleton J, Miller JA et al (2004) RNAi-mediated replacement of morphine with the nonnarcotic alkaloid reticuline in opium poppy. Nat Biotechnol 22:1559–1566

Anekonda TS (2006) Resveratrol a boom for treating Alzheimer's disease? Brain Res Rev 52:316–326

Augustin JM, Kuzina V, Anderesen SB, Bak S (2011) Molecular activities, biosynthesis and evolution of triterpenoid saponins. Phytochemistry 72:435–457

Bai W, Chern M, Ruan D, Canlas PE, Sze-To WH, Ronald PC (2011) Enhanced disease resistance and hypersensitivity to BTH by introduction of an NH1/OsNPR1 paralog. Plant Biotechnol J 9:205–215

Bailey JE (1991) Toward a science of metabolic engineering. Science 252:1668–1675

Batra P, Sharma AK (2013) Anti-cancer potential of flavonoids: recent trends and future perspectives. 3 Biotech 3:439–459

Bebek G, Koyuturk M, Price ND, Chance MR (2012) Network biology methods integrating biological data for translational science. Brief Bioinform 13:446–459

Bernohf A (2010) A brief review on bioactive compounds in plants. In: Bernhoft A (ed) Bioactive compounds in plants-benefits and risks for man and animals. The Norwegian Academy of Science and Letters, Oslo

Bishop GJ, Koncz C (2002) Brassinosteroids and plant steroid hormone signaling. Plant Cell 14:S97–S110

Bourgaud F, Gravot A, Milesi S, Gonntier E (2001) Production of plant secondary metabolites: a historical perspective. Plant Sci 161:839–851

Bradamante S, Barenghi L, Villa A (2004) Cardiovascular protective effects of resveratrol. Cardiovasc Drug Rev 22:169–188

Burns J, Yokota T, Ashihara H, Lean ME, Crozier A (2002) Plant foods and herbal sources of resveratrol. J Agric Food Chem 50:3337–3340

Chae L, Lee I, Shin J, Rhee SY (2012) Towards understanding how molecular networks evolve in plants. Curr Opin Plant Biol 15:177–184

Chang AL, Wolf JJ, Smolke CD (2012) Synthetic RNA switches as a tool for temporal and spatial control over gene expression. Curr Opin Biotechnol 23:1–10

Christou P (2013) Plant genetic engineering and agricultural biotechnology 1983–2013. Trends Biotechnol 31(3):125–127

Comalada M, Ballester I, Bailón E, Sierra S, Xaus J et al (2006) Inhibition of pro-inflammatory markers in primary bone marrow-derived mouse macrophages by naturally occur- ring flavonoids: analysis of the structure-activity relation- ship. Biochem Pharmacol 72:1010–1021

Counet C, Callemien D, Collins S (2006) Chocolate and cocoa: new sources of trans- resveratrol and trans-pieced. Food Chem 98:649–657

Croteau R, Kutchan TM, Lewis NG (2000) Natural products (secondary metabolites). In: Buchanan B, Gruissm W, Jones R (eds) Biochemistry & molecular biology of plants. American Society of Plant Physiologists, Rockville MD, pp 1250–1318

Daly JW, Kaneko T, Wilham J, Garraffo HM, Spande TF, Expinosa A, Donnelly MA (2002) Bioactive alkaloids of frog skin: combinatorial bioprospecting reveals that pumiliotoxins have an arthropod source. Proc Natl Acad Sci U S A 99(22):13996–14001

De Luca V, Laflamme P (2001) The expanding universe of alkaloid biosynthesis. Curr Opin Plant Biol 4:225–233

De Luca V, St-Pierre B (2000) The cell and developmental biology of alkaloid biosynthesis. Trends Plant Sci 5:168–173

Decendit A, Mamani-Matsuda M, Aumont V, Waffo-Teguo P, Moynet D et al (2013) Malvidin-3-O-b glucoside, major grape anthocyanin, inhibits human macrophage-derived inflammatory mediators and decreases clinical scores in arthritic rats. Biochem Pharmacol 86:1461–1467

Delgado-Zamarreño MM, Bustamante-Rangel M, Martínez-Pelarda D, Carabías-Martínez R (2009) Analysis of β-sitosterol in seeds and nuts using pressurized liquid extraction and liquid chromatography. Anal Sci 25:765–768

Dixon RA (2001) Natural products and plant disease resistance. Nature 411:843–847

Dixon RA, Chenggang L, Jun JH (2013) Metabolic engineering of anthocyanins and condensed tannins in plants. Curr Opin Biotechnol 24:329–335

Donnez D, Jeandet P, Clément C, Courot E (2006) Bioproduction of resveratrol and stilbene derivatives by plant cells and microorganisms. Trends Biotechnol 27(12):706–713

Dudareva N, Dellapenna D (2013) Plant metabolic engineering: future prospects and challenges. Curr Opin Biotechnol 24:226–228

Facchini PJ (2001) ALKALOID BIOSYNTHESIS IN PLANTS. Biochemistry, cell biology, molecular regulation, and metabolic engineering applications. Annu Rev Plant Physiol Plant Mol Biol 52:29–66

Fang SH, Rao YK, Tzeng YM (2005) Inhibitory effects of flavonol glycosides from *Cinnamomum* osmo- phloeum on inflammatory mediators in LPS/IFN-gamma-activated murine macrophages. Bioorg Med Chem 13:2381–2388

Farhi M, Marhevka E, Ben-Ari J, Algamas-Dimantov A, Liang ZB, Zeevi V et al (2011) Generation of the potent anti-malarial drug artemisinin in tobacco. Nat Biotechnol 29:1072–1074

García-Llatas G, Rodríguez-Estrada MT (2011) Current and new insights on phytosterol oxides in plant sterol-enriched food. Chem Phys Lipids 164:607–624

Geu-Flores F, Olsen CE, Halkier BA (2009a) Towards engineering glucosinolates into noncruciferous plants. Planta 229:261–270

Geu-Flores F, Nielsen MT, Nafisi M, Møldrup ME, Olsen CE et al (2009b) Glucosinolate engineering identifies a gamma-glutamyl peptidase. Nat Chem Biol 5:575–577

Giddings LA, Liscombe DK, Hamilton JP, Childs KL, Dellapenna D, Buell CR, O'Connor SE (2011) A stereoselective hydroxylation step of alkaloid biosynthesis by a unique cytochrome P450 in *Catharanthus roseus*. J Biol Chem 286:16751–16757

Glenn WS, Nims E, O'Connor SE (2011) Reengineering a tryptophan halogenase to chlorinate a direct alkaloid precursor. J Am Chem Soc 133:19346–19349

Glenn WS, Runguphan W, O'Connor SE (2013) Recent progress in the metabolic engineering of alkaloids in plant systems. Curr Opin Biotechnol 24:354–365

Goossens A, Hakkinen ST, Laakso I, Oksman-Caldentey KM, Inze D (2003) Secretion of secondary metabolites by ATP-binding cassette transporters in plant cell suspension cultures. Plant Physiol 131:1161–1164

Hagel JM, Facchini PJ (2010) Dioxygenases catalyze the O-demethylation steps of morphine biosynthesis in opium poppy. Nat Chem Biol 6:273–275

Harborne JB (1999) Classes and functions of secondary products. In: Walton NJ, Brown DE (eds) Chemicals from plants, perspectives on secondary plant products. Imperial College Press, London, UK, pp 1–25

Hawkins KM, Smolke CD (2008) Production of benzylisoquinoline alkaloids in *Saccharomyces cerevisiae*. Nat Chem Biol 4:564–573

Heim M, Frank O, Kampmann G, Sochocky N, Pennimpede T et al (2004) The phytoestrogen genistein enhances osteogenesis and represses adipogenic differentiation of human primary bone marrow stromal cells. Endocrinology 145:848–859

Heinig U, Gutensohn M, Dudareva N, Aharoni A (2013) The challenges of cellular compartmentalization in plant metabolic engineering. Curr Opin Biotechnol 24:239–246

Hernández WI, Alegre L, Van Breusegem F, Munné-Bosch S (2009) How relevant are flavonoids as antioxidants in plants? Trends Plant Sci 14(3):125–132

Hong GJ, Xue XY, Wang LJ, Chen XY (2012) Arabidopsis MYC2 interacts with DELLA proteins in regulating sesquiterpene synthase gene expression. Plant Cell 24:2635–2648

Jirschitzka J, Mattern DJ, Gershenzon J, D'Auria JC (2013) Learning from nature: new approaches to the metabolic engineering of plant defense pathways. Curr Opin Biotechnol 24:320–328

Kliebenstein D (2014) Synthetic biology of metabolism: using natural variation to reverse engineer systems. Curr Opin Plant Biol 19:20–26

Kundu JK, Surh YJ (2008) Cancer chemopreventive and therapeutic potential of resveratrol: mechanistic perspectives. Cancer Lett 269:243–261

Lee H, Kang R, Kim YS, Chung SI, Yoon Y (2010) Platycodin D inhibits adipogenesis of 3T3-L1 cells by modulating Kruppel-like factor 2 and peroxisome proliferator-activated receptor gamma. Phytother Res 24(Suppl 2):S161–S167

Liscombe DK, Usera AR, O'Connor SE (2010) Homolog of tocopherol C methyltransferases catalyzes N methylation in anticancer alkaloid biosynthesis. Proc Natl Acad Sci U S A 107:18793–18798

Misra P, Pandey A, Tiwari M, Chandrashekar K, Sidhu OP et al (2010) Modulation of transcriptome and metabolome of tobacco by *Arabidopsis* transcription factor, AtMYB12, leads to insect resistance. Plant Physiol 152:2258–2268

Moldrup ME, Geu-Flores F, De Vos M, Olsen CE, Sun J et al (2012) Engineering of benzylglucosinolate in tobacco provides proof-of-concept for dead-end trap crops genetically modified to attract *Plutella xylostella* (diamondback moth). Plant Biotechnol J 10:435–442

Naqvi S, Zhu C, Farre G, Ramessar K, Bassie L et al (2009) Transgenic multivitamin corn through biofortification of endosperm with three vitamins representing three distinct metabolic pathways. Proc Natl Acad Sci U S A 106:7762–7767

Naqvi S, Farré G, Sanahuja G, Capell T, Zhu C, Christou P (2010) When more is better: multigene engineering in plants. Trends Plant Sci 15:48–56

Noble D (2008) The music of life, biology beyond genes. Oxford University Press, Oxford, New York. ISBN-10: 0199228361, ISBN-13: 978–0199228362

Nour-Eldin H, Halkier B (2009) Piecing together the transport pathway of aliphatic glucosinolates. Phytochem Rev 8:53–67

Nour-Eldin H, Halkier B (2013) The emerging field of transport engineering of plant specialized metabolites. Curr Opin Biotechnol 24:263–270

Pang JL, Ricupero DA, Huang S, Fatma N, Singh DP et al (2006) Differential activity of kaempferol and quercetin in attenuating tumor necrosis factor receptor family signaling in bone cells. Biochem Pharmacol 71:818–826

Park SU, Yu M, Facchini PJ (2003) Modulation of berberine bridge enzyme levels in transgenic root cultures of California poppy alters the accumulation of benzophenathridine alkaloids. Plant Mol Biol 51:153–164

Patra B, Schlutternhofer C, Wu Y, Pattanaik S, Yuan L (2013) Transcriptional regulation of secondary metabolite biosynthesis in plants. Biochim Biophys Acta 1829:1236–1247

Pearson KJ, Baur JA, Lewis KN, Peshkin L, Price NL, Labinskyy N et al (2008) Resveratrol delays age-related deterioration and mimics transcriptional aspects of dietary restriction without extending life span. Cell Metab 8:157–168

Petinatti Pavarini D, Petinatti Pavarini S, Niehues M, Peporine Lopes N (2012) Exogenus influences on plant secondary metabolite levels. Animal Feed Sci Technol 176:5–16

Pilsakova L, Riecansky I, Jagla F (2010) The physiological actions of isoflavone phytoestrogens. Physiol Res 59:651–664

Piñol MT, Palazón J (1993) Metabolismo secundario. In: Azcon-Bieto J, Talón M (eds) Fisiología y Bioquímica Vegetal. Editorial Interamericana McGraw Hill, España

Purnick PE, Weiss R (2009) The second wave of synthetic biology: from modules to systems. Nat Rev Mol Cell Biol 10:410–422

Redillas MCFR, Jeong JS, Kim YS, Jung H, Bang SW et al (2012) The overexpression of OsNAC9 alters the root architecture of rice plants enhancing drought resistance and grain yield under field conditions. Plant Biotechnol J 10:792–805

Rios-Estepa R, Turner GW, Lee JM, Croteau RB, Lange BM (2008) A systems biology approach identifies the biochemical mechanisms regulating monoterpenoid essential oil composition in peppermint. Proc Natl Acad Sci U S A 105:2818–2823

Rohwer JM (2012) Kinetic modeling of plant metabolic pathways. J Exp Bot 63:2275–2292

Runguphan W, O'Connor SE (2009) Metabolic reprogramming of periwinkle plant culture. Nat Chem Biol 5:151–153

Runguphan W, Maresh JJ, O'Connor SE (2009) Silencing of tryptamine biosynthesis for production of nonnatural alkaloids in plant cultures. Proc Natl Acad Sci U S A 106:13673–13678

Sabater-Jara AB, Pedreño MA (2013) Use of β-cyclodextrins to enhance phytosterol production in cell suspension cultures of carrot (*Daucus carota* L.). Plant Cell Tissue Organ Cult 114: 249–258

Sagt CMJ (2013) Systems metabolic engineering in an industrial setting. Appl Microbiol Biotechnol 97:2319–2326

Samac DA, Tesfaye M, Dornbusch M, Saruul P, Temple SJ (2004) A comparison of constitutive promoters for expression of transgenes in alfalfa (Medicago sativa). Transgenic Research 13:349–361

Schellenberger J, Que R, Fleming RM, Thiele I, Orth JD, Feist AM et al (2011) Quantitative prediction of cellular metabolism with constraint based models: the COBRA Toolbox v2.0. Nat Protoc 6:1290–1307

Schilling T, Ebert R, Raaijmakers N, Schütze N, Jakob F (2014) Effects of phytoestrogens and other plant-derived compounds on mesenchymal stem cells, bone maintenance and regeneration. J Steroid Biochem Mol Biol 139:252–261

Shachar-Hill Y (2013) Metabolic network flux analysis for engineering plant systems. Curr Opin Biotechnol 24:247–255

Shanks J, Morgan J (1999) Plant hairy root culture. Curr Opin Biotechnol 10:151–155

Sharma S, Chopra K, Kulkarni SK (2007) Effect of insulin and its combination with resveratrol or curcumin in attenuation of diabetic neuropathic pain: participation of nitric oxide F TNF-alpha. Phytother Res 21:278–283

Song WO, Chun OK, Hwang I, Shin HS, Kim BG et al (2007) Soy isoflavones as safe functional ingredients. J Med Food 10:571–580. J Steroid Biochem Mol Biol 139: 252–261

Suttipanta N, Pattanaik S, Kulshrestha M, Patra B, Singh SK, Yuan L (2011) The transcription factor CrWRKY1 positively regulates the terpenoid indole alkaloid biosynthesis in *Catharanthus roseus*. Plant Physiol 157:2081–2093

Verma P, Mathur AK, Srivastava A, Mathur A (2012) Emerging trends in research on spatial and temporal organization of terpenoid indole alkaloid pathway in *Catharanthus roseus*: a literature update. Protoplasma 249:255–268

Verpoorte R, Memelick J (2002) Engineering secondary metabolite production in plants. Curr Opin Biotechnol 13:181–187

Vidal M, Cusick ME, Barabasi AL (2011) Interactome networks and human disease. Cell 144: 986–998

Vom Endt D, Kijne JW, Memelink J (2002) Transcription factors controlling plant secondary metabolism: what regulates the regulators? Phytochemistry 61:107–114

Winzer T, Gazda V, He Z, Kaminski F, Kern M et al (2012) A *Papaver somniferum* 10-gene cluster for synthesis of the anticancer alkaloid noscapine. Science 336:1704–1708

Woyengo TA, Ramprasath VR, Jones PJH (2009) Anticancer effects of phytosterols. Eur J Clin Nutr 63:813–820

Ye XD, Al-Babili S, Kloti A, Zhang J, Lucca P, Beyer P, Potrykus I (2000) Engineering the provitamin A (beta-carotene) biosynthetic pathway into (carotenoid-free) rice endosperm. Science 287:303–305

Yonekura-Sakakibara K, Fukushima A, Saito K (2013) Transcriptome data modeling for targeted plant metabolic engineering. Curr Opin Biotechnol 24:285–290

Zhou H, Yang X, Wang N, Zhang Y, Cai G (2007) Tigogenin inhibits adipocytic differentiation and induces osteoblastic differentiation in mouse bone marrow stromal cells. Mol Cell Endocrinol 270:17–22

Chapter 4
In Vitro Plant Cultures as Biofactories

Abstract *In vitro* plant cell cultures are defined as a culture in aseptic conditions of any part of the plant body in a nutritive media. Cells, differentiated and undifferentiated (calli and suspended cells) tissues, and organs can be maintained *in vitro*.

Some of the advantages of *in vitro* cultures are their development and maintenance in defined conditions (nutrients, light, humidity, temperature), independent from environmental influences (pests, diseases, extreme weather, geographical location, etc.), and manipulated by experienced operators in confined facilities.

Since its development, *in vitro* plant cell cultures have been an essential tool for studying plant metabolism, and to micropropagate plant species, maintaining their genetic background. Also, they were helpful to analyze and optimize the production of plant secondary metabolites.

Secondary metabolites productive processes are conducted, with a different degree of success, in undifferentiated cultures and in cultures of transformed organs. Several strategies were displayed in order to increase yields as manipulation of the plant growth regulator balance, addition of precursors, biotransformation, and immobilization.

Also, as plant cells have a different behavior in liquid culture compared with microorganisms and mammal cells, bioreactor design was adapted to develop plant cell cultures.

Keywords *In vitro* plant cell cultures-plan growth regulators • Culture media-elicitation • Undifferentiated cultures-hairy roots-scaling up

4.1 Introduction

In vitro plant cultures can be defined as the culture, in aseptic conditions, of cells, tissues, organs, and plants under definite physical and chemical conditions, in a vessel. The first report of an *in vitro* plant cell culture dates from the early twentieth century, when Haberlandt cultured single palisade cells from leaves in Knop's salt solution plus sucrose (Haberlandt 1902).

The first commercial interest on plant tissue cultures was to propagate ornamental and food species (Pacheco et al. 2012; Gonçalves and Romano 2013; Rout et al. 2000; Mulder-Krieger et al. 1988). By the end of the 1970s, on the basis of the

M.A. Alvarez, *Plant Biotechnology for Health: From Secondary Metabolites to Molecular Farming*, DOI 10.1007/978-3-319-05771-2_4

previous experience with microbial fermentations, the interest on *in vitro* plant cell cultures for producing plant metabolites escalated.

Since the 1980s, *in vitro* cultures became an important tool in basic and applied studies generating knowledge on plant metabolism and on the optimization of productive processes of metabolites in plant cells (phytofermentations).

In a first period, research was mainly focused on the production of plant natural products by *in vitro* cultures of a specific species. In a second period, the advances in genetic engineering, bioengineering, plant physiology, and the development of new areas, as metabolic engineering and molecular farming, increased the interest on plant tissue cultures for producing novel or heterologous compounds (e.g.: recombinant proteins).

Chapter 6 of this book describes the application of *in vitro* cultures for molecular farming, which is the production of heterologous proteins of commercial interest in GMOs as transgenic plants. In this chapter the focus will be on the contribution of *in vitro* plant cell cultures to the production of secondary metabolites.

4.2 *In Vitro* Cultures for Producing Secondary Metabolites

Chapter 3 covers plant secondary metabolism and the biological activities secondary metabolites display as their participation in defence against herbivores (insects, molluscs, vertebrates), and microorganisms (virus, bacteria, fungi), and their metabolic and ecological functions (attracting pollinators, animals that spread seeds, UV protection) (Wink 1988; Wink et al. 2005).

In vitro cultures are feasible due to the totipotency of plant cells, which is the capability of most of plant cells to regenerate a whole plant, identical to the parental plant, when exposed to the adequate conditions. Totipotentiality is also physiologically, which means that plant cells bear the essential genetic information for producing the same secondary metabolite pattern of the source plant.

In vitro plant cell cultures have been successfully used for producing natural secondary metabolites at laboratory scale (Table 4.1). The main advantages of *in vitro* cultures are the development of cultures in environmental controlled conditions, independent of the weather, quality of the soil, seasonality, and day-length, free from mycotoxins, herbicides or pesticides, and in relatively small facilities (Ramachandra Rao and Ravishankar 2002).

Basically, *in vitro* plant cell cultures are developed placing a piece of plant tissue, the explant (leave, stem, etc.), previously sterilized, on the surface of a nutritive medium contained in a vessel (Petri dish, flask, etc.). The nutritive medium usually contains macronutrients, micronutrients, vitamins, myo-inositol, a source of carbon (e.g.: sucrose), plant growth regulators and, if it is a semi-solid culture, a solidifying agent, usually agar. Several culture media were developed (Murashige and Skoog 1962; Gamborg et al. 1968; Schenk and Hildebrandt 1972), which are the base of almost all the culture mediums in use. Each plant species and even each duple plant-metabolite needs a culture medium with a specific composition.

Table 4.1 Secondary metabolites from *in vitro* plant cell cultures

Plant species	Secondary metabolite	Activity	Reference
Lavandula viridis Her	3-O-caffeoylquinic, 4-O-caffeoylquinic, 5-O-caffeoylquinic, rosmarinic acids, luteolin and pinocembrin	Antioxidant, anti-cholinesterase	Costa et al. (2013)
Ocium basilicum	β-caryophyllene, essential oils	Antiviral, antioxidant, antibacterial. Aroma	Bertoli et al. (2013)
Cistus creticus subsp. *creticus*	Labdane diterpenes	Antibacterial, cytotoxic	Skoric et al. (2012)
Valeriana glechomifolia	Valepotriates	Mild sedative	Russowski et al. (2013)
Larrea divaricata	Nordihydroguaiaretic acid	Antioxidant, inhibitors of lipoxygenases	Palacio et al. (2012)
Vitis vinifera	Anthocyanins	Antioxidant	Nagamori et al. (2001)
Salvia spp.	Terpenoids, polyhenols	Antioxidant, antimicrobial, anti-inflammatory, anti-carcinogenic	Marchev et al. (2013)
Berberis buxifolia Lam	Berberine	Anti-bacterial, anti-diarrheal, antispasmodic	Alvarez et al. (2009)
Silybium marianum (L.)	Silymarin	Anti-inflammatory, antioxidant, anti-carcinogenic	Sánchez-Sampedro et al. (2008)
Stevia rebaudiana Bertoni	Steviosides	Sweetener	Rajasekaran et al. (2008)
Hypericum perforatum L.	Hypericin, pseudohypericin and hyperforin	Treatment of neurological disorders and depression	Zobayed et al. (2003)
Glycyrrhiza glabra	Flavonoids	Anti-microbial	Li et al. (1998)
Panax ginseng	Ginsenosides	Anti-oxidant	Huang et al. (2013a), Huang et al. (2013b)
Thalictrum minus	Berberine	Anti-bacterial, anti-diarrheal, antispasmodic	Kobayashi et al. (1988)
Papaver somniferum	Sanguinarine	Anti-carcinogenic, antimicrobial	Park et al. (1990)

Cultures are then placed in favorable conditions (temperature, light, photoperiod, irradiance, humidity) for growth and for producing the metabolite of interest. That requires a deep analysis and optimization of the best conditions for optimal growth and production. Sometimes cultures are developed in two-steps, in the first step in a condition that favors growth and in a second step in a condition that favors metabolite production (Alvarez et al. 2009).

In a conventional approach, the development of *in vitro* plant cell cultures for producing secondary metabolites has two main stages. The first one is related to the establishment and maintenance of the *in vitro* culture. This phase involves the optimization of the conditions of culture (culture media composition, light, darkness, temperature, etc.), and the maintenance of the culture viability during the successive subcultures to fresh media. Usually, during this first step emerges the somaclonal variation, characteristic of *in vitro* cultures that imply a collection of genetic and epigenetic changes, frequently in a higher frequency than those expected in nature. That variation appears as soon as the tissues are separated from the parental plant and placed in culture. The cell proliferation produces a heterogeneous cell population, at a morphological, genetic, molecular and biochemical level, and increases with the time in culture. Thus, it is essential to perform a constant screening and selection of the producing lines (Wang and Wang 2012). If this aspect of *in vitro* cultures is neglected, the attempts to maintain high stable yields are often frustrated. The second stage is the optimization of the conditions for increasing yields and the scaling up to a bioreactor at laboratory and pilot plant scale.

4.3 Establishment and Maintenance of *In Vitro* Cultures

The selection of the plant material for starting *in vitro* cultures is the first step of the process. Explants can be any part of the plant body but in general the most used are pieces of leaves and stems, as they provide the highest amount of biomass (Nigra et al. 1987). They are heterogeneous and contain different cell types, which produce a heterogeneous cell population that differ in rates of growth, ploidy level, physiology, etc. It was demonstrated that the biosynthetic capacity of the explants source influences the yield of the *in vitro* culture. High producing parental plants often originated high producing *in vitro* cell cultures. However, variations were demonstrated among cultures from the same parental plant, e.g., *Ruta graveolens* stem-derived explants produced stem specific essential oils while those derived from roots produce root specific essential oils. Hence, the isoenzyme pattern varies among suspension cultures derived from root, cotyledon and hypocotyl (Arnison and Boll 1975). So, in some cases, the parental plant can have a significant influence on the variation of the biosynthetic activity of cultures through the heterogeneity of the explants. In other cases, the metabolite production is not related to the explant.

Once established the best starting material (explant), the focus is the establishment and maintenance of the undifferentiated cultures (calli) through the optimization of the culture conditions. The classical aspects to analyze are the environmental conditions: composition of the culture medium (inorganic compounds, iron source, vitamins, carbon and nitrogen sources, plant growth regulators), and variables as illumination, temperature, etc. The different variables are tested following a statistical design (Mize and Chun 1988).

Growth is determined by standard techniques (fresh weight, dry weight, index growth) (Ryu et al. 1990), and metabolite content with specific analytical methodologies (colorimetric, HPLC, GC, ELISA, etc.).

Once established, calli are maintained in the optimized conditions and subcultured into fresh culture medium periodically, the optimal time for transfer varies in each case. In general, samples are taken in each subculture and used to determine the variables of growth and metabolite production. It is said that different chemotypes co-habit in the tissue (Sivanandhan et al. 2013), so repetitive selection of the most productive cells is fundamental to avoid the overgrowth of less- producing cells and for establishing productive cultures (Georgiev et al. 2006; Wallaart et al. 2000).

Once performed the screening and selection of the productive lines, calli are employed for starting suspension cell cultures, which are the most convenient systems for conducting a productive process at large-scale. Cell suspension cultures are started transferring friable calli to Erlenmeyer flasks containing a culture medium of the same composition of the solid culture medium but without the gelling agent. In general, in this step factors as inoculum age and size, and concentration of some culture media components (e.g.: plant growth regulators, precursors, etc.) have to be optimized. Then, the cell suspension cultures have to be characterized in their parameters of growth and metabolite production. Growth can be measured by fresh and dry weight, packed cell volume, cell number, cell viability, medium conductivity, and/or protein and/or DNA content (Madhusudhan et al. 1995; Grant and Fuller 1971; Yeoman and Mitchell 1970; Lowry et al. 1951). Metabolites are measured as in calli.

4.4 Strategies to Improve Secondary Metabolite Production

In general, secondary metabolite yield in *in vitro* cultures is low. After establishing the profile of growth and production of the metabolite of interest, the following phase is the optimization of the conditions for achieving the highest yields. Various physical and chemical factors influences *in vitro* cultures, as pH, temperature, aeration, agitation, light, composition of the culture medium including plant growth regulators, addition of precursors and/or elicitors, systems of culture (immobilized cells, undifferentiated or differentiated cultures), process operation (batch, fed-batch, continuous culture), etc. (Benlarbi et al. 2014; Malik et al. 2011).

4.4.1 Age and Inoculum Size

The age of the inoculum is a key factor as the production of the secondary metabolite is sensitive to the physiological age of the inoculated cells (Parsons et al. 2001). Also, the size of the inoculum is significant, the amount of cells that are inoculated in fresh media in every sub-culture affects the rate of growth as well as the secondary metabolite accumulation rate and final yield (Parsons et al. 2001; Zenk et al. 1977). In general, lowering the density of cells inoculated increases the specific rate of growth and the speed of synthesis and decreases the final yield.

4.4.2 Culture Media Composition

The media composition (carbon, nitrogen, phosphorus and potassium sources) affects both the primary as the secondary metabolism (Nigra et al. 1989; Shinde et al. 2009; Karwasara and Dixit 2012).

Essential elements of culture media are the macroelements, nitrogen, phosphorus, potassium, calcium, magnesium and sulphurum; the microelements iron, manganese, zinc, boron, copper, molybdenum, a carbon source (sucrose, glucose, lactose, maltose, starch); vitamins, myo-inositol and PGR.

It was established that, in general, there is a reverse relationship between cell growth and product yield with a tendency of the secondary metabolites to accumulate in the later phase of growth. Thus, one of the proposed strategies is restraining growth by limiting the carbon, nitrogen or phosphorus sources.

Other components of the culture media can also be modified in order to increase yields. For example, cell suspension cultures of *Silybum marianum* (L.) Gaertn were grown in a Murashige and Skoog calcium- free culture medium resulting in an increase of silymarin accumulation. A paradoxical case is that the production of isoflavonoids by soybean cultures decreased by removing calcium from the culture medium (Stäb and Ebel 1987).

Table 4.2 shows the components of the Murashige and Skoog medium, which is one of the most used in *in vitro* cultures.

4.4.3 Plant Growth Regulators (PGR)

PGR have a remarkable influence on cell growth and secondary metabolite. While primary metabolism brings the energy and the structural units for plant life, PGR regulate the rate of growth of each individual part and integrates them in the

Table 4.2 Murashige and Skoog media components (Murashige and Skoog 1962)

Group	Components
Major inorganic nutrients	NH_4NO_3, KNO_3, $CaCl_2.2H_2O$, $MgSO_4.7H_2O$, KH_2PO_4
Trace elements	KI, H_3BO_3, $MnSO_4.4H_2O$, $ZnSO_4.7H_2O$, $Na_2MoO_4.2H_2O$, $CuSO_4.5H_2O$, $CoCl_2.6H_2O$
Iron source	$FeSO_4.7H_2O$, $Na_2EDTA.2H_2O$
Vitamins	Myo-inositol, nicotinic acid, pyridoxine-HCl, thiamine-HCl, glycine
Carbon source	Sucrose

whole plant. In general, PGR are plant-organic compounds that influence physiological processes at low concentrations. Those processes are growth, differentiation and development. Other processes, as stomata movement, can be affected by PGR as well. PGR are in general into one of the following groups: auxins, cytokinins, gibberellins, ethylene, abscisic acid, and jasmonic acid. From them, auxins and cytokinins play an important role in *in vitro* cultures. It is well known that the relationship auxins: cytokinins induce calli development and also plant regeneration from calli. Additionally, they have a great influence on secondary metabolite production (Muthaiya et al. 2013).

The prototype of auxins is the indole acetic-3-acid (IAA), which is a natural PGR that can be found conjugated to aspartic acid. IAA precursors can also have auxin activity (e.g.: indole acetic aldehyde), and also other components as phenyl acetic acid.

In general, auxins promote cell growth and elongation. In *in vitro* cultures, they induce cell division, cell elongation, formation of adventitious roots, and inhibition of adventitious and axillary shoot formation, induction of embryogenesis, callus initiation and growth. Synthetic auxins as 2,4- dichlorophenoxyacetic acid (2,4-D), naphthalene acetic acid (NAA) and indole butyric acid (IBA) are frequently used for their chemical stability (Morini et al. 2000).

Other compounds in use that are not chemical related to auxins but display the same activity are 3,6-dochloro-o-anysic acid (DICAMBA), 4-amino-3,5,6-trichloropiridine-2-carboxylic acid (PICLORAM) which at high concentrations are herbicides. There are a great number of reports related to the influence of auxins on calli initiation (Nigra et al. 1987; Ahmad et al. 2010).

Sometimes, *in vitro* cultures loss, in an inheritable form, their need of auxins, a phenomenon known as habituation (Kevers et al. 1996).

Auxins are generally used at a concentration range of 0.01–10.0 mg l^{-1}.

Cytokinins (CK) derive from adenine and can be present as a riboside or ribotide. They are characterized for their ability of inducing cell division in combination with auxins. The most common natural cytokinin is zeatine. In *in vitro* cultures the most used are kinetine and zeatine (Aremu et al. 2013). Also benzyl-aminopurine (BAP), a synthetic cytokinin, is used. BAP induces adventitious shoots in *Larrea divaricata* and *Echinacea angustifolia* D.C. (Palacios et al. 2012; Lucchesini et al. 2009).

Cytoquinins regulate cell division, stimulate auxiliary and adventitious shoot proliferation, differentiation, and inhibit root formation. Also, they activate RNA synthesis.

In general cytokinins are used in *in vitro* cultures in the presence of auxins. When only a cytoquinin is needed it is assumed that the cells are producing their own auxins (Flick et al. 1983).

In *in vitro* cultures cytoquinins are used at a concentration range of 0.01–10.0 mg l^{-1}.

The abscisic acid (ABA) is a PGR synthesized from mevalonic acid in mature leaves, mainly in response to hydric stress. Seeds are also rich in ABA. In somatic embryogenesis, ABA can prevent early germination allowing a better embryonic development. It was demonstrated that ABA participates in the response to abiotic stress through the stimulus of stomatal closure, desiccation tolerance and dormancy induction in seeds. ABA is occasionally used to postpone growth and to moderate the effect of auxins and cytokinins. In plant tissue cultures ABA is used as inhibitor of plant growth for slow-growth conservation, anti-transpirant during plant acclimatization and to induce maturation or the quiescent state in somatic embryos, among other activities (Rai et al. 2011; Prakash and Gurumurthi 2010; Vahdati et al. 2008; Shiota and Kamada 2008). Also, ABA displays a protective effect against cytokinin-induced programmed cell death (apoptosis) (Carimi et al. 2003). ABA synthesis is inhibited by fluridone.

Gibberellins (GA) are a family of compounds (more than 90) based in the structure of the *ent*-giberellane. The most widely distributed GA is the gibberelic acid (GA3), which is produced by fungi. In plants the most significant is GA1 responsible of stem elongation. However, GA utilization in *in vitro* cultures is not widely distributed. It was demonstrated that the addition of GA3 affects calli expansion, viability, stem and roots regeneration, etc. On the contrary, in cell suspension cultures sensibility to GA3 is related to the number of cell divisions after the stationary phase, the duration of the lag phase in cultures with low inocula density, and to the secretion of amilase, peroxidase, phenolic compounds and pectinic acid. In *in vitro* cultures of *Datura inoxia* and *Brugmansia candida* hairy roots, GA3 promotes the increase of fresh weight, elongation and formation of lateral branches. Although cell suspension cultures are generally not responsive to GA3, certain species of *Spinacia* and *Rosa* are sensitive (Randoux et al. 2012; Dalton and Street 1976). Paclobutrazol, ancymidol, and other similar compounds inhibit gibberellins synthesis.

It is not possible to generalize about the type and quantity of PGR adequate for developing a productive *in vitro* culture; a case-by-case analysis has to be performed. However, there is an immense literature on the subject. Some examples are the production of vincristine and vinblastine by *C. roseus* (Scragg et al. 1989; Zhao et al. 2009; Taha et al. 2009; Verma et al. 2013), ajmalicine by *Rauwolfia serpentina* (Zenk et al. 1977; Goel et al. 2009; Susila et al. 2013), berberine by *Berberis buxifolia* or *Thalictrum* sp. (Alvarez et al. 2009; Smolko and Peretti 1994; Kim et al. 1990; Suzuki et al. 1988; Nakagawa et al. 1984), etc.

Other strategies to increase secondary metabolite production have also been effective (Sharma et al. 2011) as the removal and addition of precursors, differentiation and compartmentation, elicitation, immobilization, biotransformation, and transformed cultures.

4.4.4 Removal and Addition or Precursors

The removal of end products is attempted to eliminate its inhibitory effect. The beneficial effect of this approach was demonstrated by the increase in ajmalicine content following the addition of a resin enclosed in porous Miracloth to *C. roseus* suspensions (Wong et al. 2004). The addition of precursors or intermediates, whose

Table 4.3 Productivity of secondary metabolites of commercial interest produced in plant *cell in vitro* culture

Secondary metabolite	Cell line	Productivity (g l^{-1} d^{-1})	Reference
Anthocyanin	*Vitis* (Bailey Alicant A), *Vitis* (hybrid)	0.06	Yamakawa et al. (1983)
Berberine	*Thalictrum minus*	0.05	Kobayashi et al. (1988)
	Coptis japonica	0.60	Fujita (1988)
		0.1	Hussain et al. (2012)
Diosgenine	*Dioscorea* spp.	0.75	Fowler and Scragg (1988)
Rosmarinic acid	*Coleus blumei*	0.91	Ulbrich et al. (1985)
Sanguinarine	*Papaver somniferum*	0.034	Park et al. (1992)
Shikonine	*Lithospermum erythrorizon*	0.15	Tabata and Fujita (1985)

influence diminishes with the number of intermediate steps between them and the end product, have been effective in certain cases. In some situations, supplementing endogen or exogen precursors can increase the *in vivo* enzymatic activity. In most of the *in vitro* cultures, however, even in optimal conditions, the biosynthetic rate of secondary metabolites remains low due to the limited expression of the genes for the enzymes of the secondary metabolism (Table 4.3).

The production of the cytotoxic lignan nordihydroguaiaretic acid (NDGA) and the phenylpropanoids p-coumaric acid, ferulic acid and sinapyl alcohol, and cell growth were analyzed in *L. divaricata* Cav. plant cell cultures after the addition of the precursors L-phenylalanine, cinnamic acid, ferulic acid, and sinapic acid. Only the addition of L- phenylalanine (0.5, 1.0, 3.0 mM) resulted in an increase of NDGA and p-coumaric acid. On the contrary, increasing levels of L-phenylalanine resulted in a diminution of the bioconversion into sinapyl alcohol. Cinnamic acid at low concentrations (0.5 mM) only stimulated growth, whereas cinnamic acid at higher concentrations (1.0 and 1.5 mM) and ferulic and sinapic acid at all the concentrations tested resulted toxic to the cell (Palacio et al. 2011).

4.4.5 *Elicitation*

The addition of elicitors, triggers the accumulation of phytoalexines (Patel and Krishnamurthy 2013; Cusido et al. 2014). Elicitors can be abiotic (heavy metal salts, UV radiation, chemicals with high affinity to DNA or that disrupt the cell membrane) or biotic (structural polysaccharides of e.g.: fungi mycelia wall) (Savitha et al. 2006; Antognoni et al. 2007; Wiktorowska et al. 2010; Cai et al. 2011; Bertoli et al. 2013).

The production of ginsenoside by suspension cultures of *Panax ginseng* was elicited by vanadate, added at the 4th day of culture at a concentration of 50 μM. In this case, the elicitation was mediated by jasmonic acid (JA) and the transcriptional levels of genes that codify for key enzymes in the biosynthesis (*sqs*, *se* and *ds*) were increased (Huang and Zhong 2013). The same research group demonstrated that the H^+-ATPase inhibitor N,N′-dicyclohexylcarbodiimide (DCCD) was an effective abiotic elicitor that triggers the NO-mediated signal transduction pathway, with an increase in the *sqs*, *se*, and *ds* gene expression, and as a consequence an enhanced production of ginsenosides (Huang et al. 2013a).

Methyl jasmonate also elicited by 5–6 times the production of isoflavonoids by *Glycine max* plant cell suspension batch culture (Gueven and Knorr 2010).

An interesting example of elicitation is the effect of salicylic acid combined with ultrasound on the production of valepotriates by *Valeriana glechomifolia* plants growing in liquid medium. Each elicitor improved independently the valepotriates content almost 2-times, the first one causing a slight inhibition of growth depending on the concentration and time of exposition to the elicitor, the second one not affecting growth in any of the conditions tested. Conversely, salicylic acid did not modify valepotriates content in *V. glechomifolia* roots, maybe because in roots the biosynthesis of valepotriates is not through a salicylic-independent pathway (Russowski et al. 2013).

The effect of ultrasound eliciting the biosynthesis of secondary metabolites in cultures in liquid medium can be explained by the induction of hydrodynamic events that cause shear stress to plant cells, which trigger a defense-responding activation of genes. Ultrasound resulted also beneficial for the accumulation of Taxol® by *Taxus chinensis* (Wu and Ge 2004).

The combined effect of JA and β-cyclodextrins (β-CD) was also demonstrated to elicit the production of phytosterols by cell suspension cultures of *Daucus carota*. The β-CD combines a promoter effect of the remotion of product from the culture medium with an inductor effect on metabolite biosynthesis, which is enhanced by JA (Sabater-Jara and Pedreño 2013).

Another elicitation strategy is the addition of fungal extracts to culture media. Pérez et al. (2001) have increased peroxidase amount in hairy root cultures of *Armoracia lapathifolia* by the addition of extracts of fungal plant pathogens (*Colletotricum graminicola, Fusarium moniliforme, Phomopsis, F. solani, N. vasinfecta, P. acanthicum, Pythium, and Phoma medicaginis* var. pinodella).

4.4.6 Differentiation and Compartmentation

The accumulation of secondary metabolites can be according two different patterns related to plant growth and development: (a) they accumulate during the stationary growth phase or associated to differentiation, as is the case of the anthocyanines produced by *V. vinifera* cultures; (b) accumulation during the exponential growth phase, as is the case of isoflavonoids by soy plant callus suspension culture (Gueven and Knorr 2010). The second case is the most frequent because plants need a certain degree of differentiation for the synthesis of determined secondary metabolites.

Although the biosynthesis of a secondary metabolite can take place in all the tissues and cells of the plant, usually the synthesis of some metabolites is restricted to specific tissues and even to specific cells (Nigra et al. 1989; Luckner 1990). That can be associated to a lack of expression of genes that control essential pathway steps in non-specialized cells, a deviation of substrates form the biochemical pathway, an inoperability of the transport mechanisms that allow the removal of potentially toxic end products, the unavailability of secondary metabolites characteristics storage sites, and the catabolic deregulation of the synthesized product. Therefore, the lack of secondary metabolites could be the consequence of an incorrect or absent cell specialization. Frequently, the differentiation of undifferentiated tissues produces a partial or complete restoration of the capacity of accumulating secondary metabolites. Although the production of secondary metabolites in regenerated organs (adventitious stem and roots) via the manipulation of the PGR relationship was broadly studied, the most remarkable results were obtained after the development of the technology of plant transformation.

Also, cell compartmentation has also to be considered. Some molecules are produced in the cytoplasm, while others are produced in chloroplasts, mitochondria, or the endoplasmic reticulum. The formation of ginsenosides in *in vitro* cells of *Panax japonicus* var. repens is proposed to be closely related to compartmentation since the accumulation of the PPD-type ginsenosides is strongly dependent on malonylation that targets them to the vacuoles (Kochkin et al. 2013).

Precursors and intermediate molecules are compartmentalized and driven to specific metabolic steps. *In vivo*, most of the enzymes from the secondary metabolism work in limiting conditions of substrate and do not achieve their maximal rate (Sakuta and Komamine 1987).

Secondary metabolites can accumulate in the place of synthesis or can be transported to other tissues or plant organs. In some cases their production is restricted to a specific development stage (e.g.: flowers, fruits, plantlets, etc.) or to ecological or environmental conditions (e.g.: phytoalexins after a wound or microbial attack).

One of the disadvantages of *in vitro* plant cell cultures is their low growth, fragility and genetic instability that demand a continual selection of producing lines. According to their productive capacity of secondary metabolites, *in vitro* plant cell cultures can be classified into four categories: (a) cultures that always produce the compound (*Morinda*, *Galium*, producers of antraquinones), (b) cultures with producing cells in presence of a majority of non-producing cells (*Catharanthus*, *Solanum*, producers of ajmalicine and solasodine respectively), (c) cultures that usually do not produce the compounds that the parental plant produces (*Papaver somniferum*, *Datura*, producers of morphine and scopolamine respectively), and (d) cultures that produce compounds that are not produced by the parental plant (Ruyter and Stöckgit 1989; Schübel et al. 1989; Zhao and Verpoorte 2007).

The disadvantage for the commercial use of cell cultures is that most of them fail to produce the compound of interest, maybe for lack of precursors, or for the activity of the key enzyme in the metabolic pathway.

4.4.7 Biotransformation

Plant cells can use its biosynthetic machinery to catalyze the biotransformation of diverse compounds like phenolics, steroids, alkaloids, etc., by a wide variety of reactions (oxidation, reduction, hydroxylation, methylation, glycosylations, esterifications, etc.). They can act as biocatalysts of foreign and even synthetic molecules if the necessary enzymes are expressed in the plant cell (Giri et al. 2001). In general, the rate of conversion substrate-product is high and the reactions are stereo- and regiospecific, being the production costs justified by the added value of the product.

Considerable improvements have been done by genetic engineering, cloning in plant cells a foreign gene to induce or improve the biosynthesis of a specific compound. Conversely, genes codifying for plant enzymes have been cloned into micro-organism to profit from their high productive rates (Alvarez and Marconi 2011).

4.4.8 Immobilization

Immobilization force plant cells to growth in close contact, in a microenvironment that brings the conditions to develop a certain degree of organization favourable to the production of the secondary metabolite. Basically, plant cells are entrapped in a polymeric matrix (alginate, agar, carragenanes, chitosan, gelatine beads or surfaces), and cultured in vessels in environmental conditions to maintain cells in the stationary phase of growth with a certain degree of structural and biochemical differentiation. The product is secreted into the culture medium and removed through classical downstream processing. The process can be developed in a semi-continuous or continuous operation.

Immobilization can also be combined with biotransformation or/and with elicitation. As an example, cell suspensions of *Vitis vinifera* and *Cruciata glabra* were immobilized in pectine/chitosan coacervate capsules and treated with chitosan or pectins as elicitors. The treatment produced an increase in the levels of hydrogen peroxide in both cases, and of anthoçyanines in *V. vinifera* and anthraquinones in *C. glabra*. The same positive effect was of lower intensity in cells growing freely suspended, demonstrating the synergistic effect of immobilization and elicitation (Dörnenburg 2004).

4.5 Plant Transformation – *Agrobacterium*

In the 1980s the mechanism by which the natural plant pathogen *Agrobacterium* produces the crown gall disease in plants was elucidated. It was established that when *Agrobacterium* enters into the plant cell after a wound, introduces a fragment of its plasmidic DNA (plasmid Ti or plasmid Ri in *A. tumefaciens* and *A. rhizogenes* respectively), which is integrated into the plant genome. The T region contains oncogenes that codify for enzymes that participate in auxin and cytokinin synthesis when expressed in plants. In the Ri plasmid, the genes codify for enzymes that determine high tissue sensitivity to PGR. The transformation with *A. rhizogenes* triggers the development of apical roots that will originate transformed roots, in the infection sites hairy roots. The transformation with certain strains of *A. tumefaciens* originates shooty teratomes. Once the transformed organ was obtained it can growth without any external supply of PGR.

The T-DNA also contains genes that codify for the synthesis of opines, used by bacteria as unique carbon and nitrogen sources, which are the base of the classification of the *Agrobacterium* strains (de la Riva et al. 1998).

4.5.1 Transformed Organs

Transformed roots obtained by infection with *A. rhizogenes* spp. (e.g.: LBA9402) are also called hairy roots for their abundant absorbent hairs. Shooty teratomes are obtained after transformation with *A. tumefaciens* strains (e.g.; T37) that contain genes for the synthesis of auxins (*aux1* and *aux2*) and cytokinins (*ipt*) that produced a reduced index auxin: cytokinin and as a consequence induces the production of stems.

Hairy roots have certain advantages respect to cell suspension cultures, they grow relatively faster, are biochemical and genetically stable, and produce a spectrum of secondary metabolites that quali- and quantitatively resembles those of the parental plants and in some cases are even more productive (Agostini et al. 2013; Ono and Tian 2011; Dörnenburg and Knorr 1995). Hairy roots allow the study of secondary metabolic pathways being relevant for the functional analysis of genes, plant physiology and metabolic research (Sharma et al. 2013; Hu and Du 2006). Also, they reunite the advantage of *in vitro* cultures with the presence of the *rol* genes that induce secondary metabolite pathways, even of those that are synthesized in aerial organs (Wheathers et al. 2005).

The genetic and biochemical stability of hairy root avoids the constant screening and selection of productive lines characteristic of undifferentiated *in vitro* cultures. The increased accumulation of certain secondary metabolites in hairy roots can be attributed to the fact that the *rolB* and *rolC* genes not only activate phytoalexin production but also suppress intracellular reactive oxygen species (ROS) levels (Bulgakov et al. 2011). Apparently, ROS mediate the elicitor-induced accumulation of isoflavonoids in species as *C. roseus*, *P. ginseng*, etc. (Zhao et al. 2005). The stimulator effect of *rolC* on growth and secondary metabolite production can be favorable in a production process. On the other hand, *rolB* that also has a positive effect on secondary metabolism is not attractive due to its inhibitor effect on growth (Bulgakov 2008).

Table 4.4 Compounds of plant origin produced in hairy root cultures

Plant species	Compound	Reference
Arachis hypogea	Stilbenes	Medina Bolívar et al. (2007)
Artemisia dubia	Artemisinin	Mannan et al. (2008)
A. indica		
A. lapathifolia	Peroxidases	Pérez et al. (2001)
Scopolia parviflora	Scopolamine	Jung et al. (2003)
Medicago trunculata	Carotenoids	Floss et al. (2008)
Taxus cuspidata	Paclitaxel	Kim et al. (2009)
Glycyrrhiza glabra	Fkavonoids	Li et al. (1998)
Astragalus membranaceus Bge	Cycloartane	Ionkova et al. (2010)
	Saponins	
D. straminium	Hyosciamine	Pavlov et al. (2009)
Arachis hypogaea	Resveratrol	Condori et al. (2010)
	Stilbenoids	
C. roseus	Ajmalicine	Wong et al. (2004)
P. ginseng	Ginsenoside	Yu et al. (2005)

The different strategies applied to undifferentiated cultures in order to improve secondary metabolism production (culture medium composition, elicitators, precursors, etc.) were adapted to hairy root cultures (Elwekeel et al. 2012; Georgiev et al. 2012; Thimmaraju et al. 2003) (Table 4.4).

The production of tropane alkaloids by hairy roots of *Datura stramonium, Hyosciamus muticus, Brugmancia candida, Atropa belladonna* is an interesting example of the employment of different strategies. Elicitation, overexpression of key enzymes, genetic transformation with homologous genes, and the use as biocatalyzers of engineered *N. tabacum* and yeast overexpressing the *h6h* DNA gene were performed with different degree of success (Alvarez and Marconi 2011).

Even though the system has its advantages, their commercial application is not widespread. The main drawback for their massive culture is the characteristics of growth of the highly branched roots. Several attempts have been done to overcome that hurdle with a relative success (Medina-Bolivar and Cramer 2004).

An interesting development related to hairy root cultures is the Semiautomatic Image Processing System, RHYZOSCAN, a nondestructive analytical method, for determination of root architecture and secondary metabolite concentrations from scanned images (Berzin et al. 2000). The software resulted suitable for pigments or stainable products, because colour image analysis is implicated, but not appropriate for thick root clusters.

Hairy roots were not only used for producing plant metabolites but also for producing recombinant proteins. Chapter 5 summarizes some of advances in that field.

Also, a compilation of patents related to the application of hairy roots with different purposes (natural product production, bioremediation, etc.) can be find in Talano et al. (2012).

4.6 Metabolic Engineering

Metabolic engineering, synthetic engineering, and system biology engineering are devoted to elucidate biosynthetic pathways and also to offer the tools needed to improve yields (Mora-Pale et al. 2013; Wilson and Roberts 2014). Nevertheless, the number of commercial *in vitro* plant cell cultures processes for producing natural products remains low.

The accumulation and storage of secondary metabolites is the consequence of complex interactions among biosynthesis, transport, storage and degradation of the metabolite. Those processes are genetically regulated and the expression of the involved genes can be modulated through bioengineering, e.g.: by over-expressing the genes for key enzymes in the metabolic pathway (Verpoorte and Memelick 2002). An example of the successful implementation of metabolic engineering is the manipulation of anthocyanins and condensed tannin metabolism by regulating a family of transcription factors (MYB). The astringency, bitterness and color of fruits, the defence against predators (fungus and herbivores), and the forage quality can be manipulated in order to generate plants with the commercial desirable traits (Dixon et al. 2013).

Multiple-gene transformation of hairy roots has also been used to overexpress or suppress genes (Peebles et al. 2011; Huang et al. 2013b). An example is the culture of transgenic *C. roseus* hairy roots; the approaches employed were two *A, rhizogenes* strains carrying each one a binary vector with the marker genes, an *A. rhizogenes* strain with both binary vectors with the marker gene, or an *A. rhizogenes* with the transgenes in the same or different T-DNA region of a single binary vector. With the last two strategies, similar co-transformation efficiency was obtained despite of the high size of the vector plus transgenes. While, when the transgenes are in different binary vectors, the co-transformation efficiency was lower, perhaps for plasmid incompatibility. However, it can be an alternative tool when trying to introduce a large number of transgene in the plant genome (Huang et al. 2013b).

An example of the application of metabolic flux characterization is the production of geraniol engineered in tobacco hairy roots through the expression of geraniol synthase from *Valeriana officinalis* with/without geranyl phosphatase synthase from *Arabidopsis thaliana*, both plastid-targeted (Masakapalli et al. 2014). The effect of over-expressing geraniol synthase resulted in an increase of geraniol, as glycoside, effect that was not improved by the combined expression of enzymes. The major fluxes of central carbon metabolism did not show differences in the transgenic lines, and were not limited by substrate supply.

4.7 Recombinant Proteins

The expression of recombinant proteins in plants is also called molecular farming. The ability of plant to express foreign proteins was demonstrated at the end of the 1980s by several groups. In Belgium, the group headed by Schell and van Montagu

(Herrera-Estrella et al. 1983) described the use of *Agrobacterium* strains as a vector for introducing heterologous genes into plants. Since then, a huge number of proteins (enzymes, antibodies, antigens, etc.) were expressed in different plant species. Chapter 6 is devoted to this subject.

4.8 Scale-Up

The necessity of scaling-up the cultures of transformed organs triggered the development of bioreactors adapted to growing organs in liquid medium. After optimizing the productive process in Erlenmeyer flasks and analyzing the feasibility of a commercial exploitation, the next step is to scaling-up to bioreactors (Lee et al. 2004). Plant cells in culture have different characteristics than mammal and microbial cells (Table 4.5), as a consequence, there are some critical points to considered in order to adapt classical bioreactors to plant cell cultures.

4.8.1 Aeration

Oxygen requirement by plant cells is relatively low but can significantly increase during metabolite synthesis. The oxygenation is the main preoccupation of bioreactor design and scaling up since it limits cell growth, and consequently the volumetric productivity of high density and high viscosity cell suspensions. The adequate oxygen demand has to be established maintaining the adequate levels of CO_2 and other essential volatile compounds. The oxygen transference coefficient (Kla) has to be established (Thanh et al. 2006; Lee et al. 2011).

Table 4.5 Characteristics of microbial, mammal, and plant cell cultures

Characteristics	Microbial cells	Mammal cells	Plant cells
Size	1–5 μ	10^5–10^6 μ	10^5–10^6 μ
Shear stress sensitivity	Low	High	High
Type of growth	Single cells	Single cells or surface adhered	Cell aggregates
Duplication time	<1 h	20 h	>24 h
Aeration	1–2 vvm	0.2 vvm	0.2 vvm
Nutritive requirements	Simple	Complex	Complex
Cell density	10^{10}cells ml^{-1}	10^6cells ml^{-1}	10^6cells ml^{-1}
Oxygen requirement	<180 mmol l^{-1} h^{-1}	0.7–1.0 mmol l^{-1} h^{-1}	0.5–1.8 mmol l^{-1} h^{-1}

4.8.2 Mixing

It is essential to have a homogenous distribution of cells, nutrients and products. Mixing is performed with aerators, mechanical stirrers or a combination of both. The magnitudes of hydrodynamic forces associated to mixing must be sufficiently low as to not cause cell death and to stimulate certain cell functions.

In mechanically agitated bioreactors, different types of impellers are used to agitate medium (flat-blade turbine impeller, helical ribbon impeller, vibrating perforated plates). Pneumatically agitated bioreactors can be bubble columns where air is bubbled at the base of the column, or air lift where air is sparged from the riser section to the top of the column and then the medium descends in the down corner section.

4.8.3 Culture Medium Selection

The selection of the operation condition (batch, fed-batch or continuous culture) depends on the dynamic of the specific culture. Batch cultures are characterized by conditions constantly changing, metabolites are produced associated to a kinetic pattern, and the culture medium is, generally, the one optimized at lab scale.

Fed-batch cultures are those with t a controlled addition of nutrients, generally a limiting one that permits the study of the conditions for maximal productivity in long periods of time.

Continuous culture maintains cells in stationary state (chemostat) with a constant supply of culture medium and cells; it is usually designed for producing a high volume of low value products associated to growth, typically primary metabolites and biomass (Bondarev et al. 2003; Antoniukas et al. 2006; Desai et al. 2006).

4.8.4 Bioreactors

In a bioreactor, the optimization implies a balance between biological and engineering factors to obtain high productivity at minimal costs. Plants cells are characterized by slower growth, lesser oxygen demand, and higher sensibility to shear stress than microbial cells, and also by the tendency to form clusters by aggregation. Thus, the culture of plant cells in bioreactors requires of specific aeration and mixing devices.

The design of the bioreactor is a key factor for providing an accurate mixing and mass transfer minimizing the intensity of shear stress and hydrodynamic pressure. Thus, the design of bioreactors involves taking into account all that characteristics of plant cells (particularly shear stress, doubling time, photoperiod, etc.).

The type of bioreactor (stirred, air lift, plastic sleeve) (Table 4.6) must be selected considering the type of operation to be conducted (batch, fed-batch, continuous),

Table 4.6 Advantages and disadvantages of bioreactors for plant cultures

Bioreactor	Advantages	Disadvantages
Stirred tanks	Flexible, high coefficient of mass transfer, homogenous, useful for high density cultures, useful for working with GMPs.	Shear stress, expensive, risks of contamination (valves), heat emission.
Pneumatics (bubble column) bioreactors	Easy scale-up, low costs, low contamination risks, low shear stress, low heat emission.	Low oxygen transfer, inefficient mixing in high viscosity cultures, foam production at high aeration.
Pneumatics (airlift)	Bubbling driving, good oxygen transfer, low shear stress, without heat emission, low mixing times, low costs, easy scale-up, efficient fluid circulation.	Inefficient mixing in high viscosity cultures, foam production at high aeration.
Wave reactor	Without shear stress, good oxygen transference, low operational costs.	Scale-up, heat emission.
Membrane-reactor	Removal of extracellular products, low shear stress, low operational costs.	Scale-up, oxygenation, low heat transference, on-line monitoring.

and the behavior of the metabolite to be produced (secreted or not to the culture media, produced associated or not to growth, etc.).

Strategies as elicitation, removal of products *in situ* to avoid feedback inhibition, etc., are also valid.

Other parameters to be optimized are the simplicity and celerity of the biomass harvest and of product extraction, purification and quantification (Archambault et al. 1996; Chaterjee et al. 1997; James et al. 2002; Komar et al. 2004; Dewir et al. 2006; Curtis 2005; Huang and Mcdonald 2009; Donnez et al. 2011; Ferri et al. 2011; Baque et al. 2012).

Batch culture is a close system where the inoculum is fixed at the beginning of the process. The conditions in the vessel change as the nutrients are consumed and end products accumulate. In general the kinetic of growth is a sigmoid curve with a lag, exponential and stationary phase, the last when the culture enters in senescence.

Fed batch culture receives periodically fresh culture medium to maintain plant biomass growth (in a linear relationship).

Continuous culture receives fresh medium during the exponential phase of growth and a similar volume of culture is retired in order to maintain the volume in the vessel.

During the whole process the fundamental parameters of culture must be monitored. Some parameters are monitored on-line (temperature, pH, dissolved oxygen, carbon dioxide, foam) while others are off-line controlled (cell concentration and viability, tissue morphology, substrate and product concentration) (Forcato et al. 2002; Nadadoor et al. 2012).

4.9 Modeling

Mathematical models are essential for understanding the behavior of cultures under different conditions. Chapter 9 is dedicated to the subject of mathematical modeling in *in vitro* cultures.

4.10 Conclusive Remarks

Only a few productive processes for producing secondary metabolites using *in vitro* cultures were commercially established, although there is a great amount of knowledge generated about biosynthetic pathways, methods for optimizing yields, large-scale operations, etc.

Up to now, large-scale processes have only been developed in a few countries (Japan, Russia. Germany). Japan has promoted this technology with strong politics of academy-industry collaboration. As a result the dye shikonine is produced by *in vitro* cultures of *L. erythrorizon* in a commercial scale (Japan Tobacco Inc's). Japan also produces alkaloids, steroids, etc. in *in vitro* cultures of *Panax ginseng* (Meiji Seiko, Ajinomoto and Nippon Shin-Jaki) (Yesil Celiktas et al. 2010; Curtin 1983).

The limited industrial applications of *in vitro* cultures for producing secondary metabolites in other countries are mainly attributed to the low yields and the loss of the productive capability with time of those cultures. Optimization through bioreactor design can be of great help for large-scale production.

New devices, (e.g. meshes to growth hairy roots), or the induction of a certain degree of organization in single cell cultures (e.g.: immobilization) can generate the conditions that cultures need to reach their maximal biosynthetic capacity.

References

Agostini E, Talano MA, González PS, Weaver Oller AL, Medina MI (2013) Application of hairy roots for phytoremediation: what makes then an interesting tool for this purpose? Appl Microbiol Biotechnol 97:1017–1030

Ahmad N, Faisal M, Anis M, Aref IM (2010) In vitro callus induction and plant regeneration from leaf explants of *Ruta graveolens* L. S Afr J Bot 76:597–600

Alvarez MA, Marconi PL (2011) Genetic transformation for metabolic engineering of tropane alkaloids. In: Alvarez MA (ed) Genetic transformation. Croatia 15. pp 291–304

Alvarez MA, Fernández Eraso N, Pitta-Alvarez S, Marconi PL (2009) Two–stage culture for producing berberine by cell suspension and shoot cultures of *Berberis buxifolia* Lam. Biotechnol Lett 31:457–463

Antognoni F, Zheng S, Pagnucco C, Baraldi R, Poli F, Biondi S (2007) Induction of flavonoid production by UV-B radiation in *Passiflora quadrangularis* callus cultures. Fitoterapia 78: 345–352

Antoniukas L, Grammel H, Reichl U (2006) Production of hantavirus *Puumala* nucleocapsid protein in *Saccharomyces cerevisiae* for vaccine and diagnostics. J Biotechnol 124:347–362
Archambault J, Williams RD, Perrier M, Chavarie C (1996) Production of sanguinarine by elicited plant cell culture III. Immobilized bioreactor cultures. J Biotechnol 46:121–129
Aremu AO, Gruz J, Subrtová M, Szucová L, Dolezal K et al (2013) Antioxidant and phenolic acid profiles of tissue cultured and acclimatized *Merwilla plumbea* plantlets in relation to the applied cytokinins. J Plant Physiol 170:1303–1308
Arnison PG, Boll WG (1975) Isoenzymes in cell cultures of bush bean, (*Phaseols vulgaris* cv, Contender); isoenzymatic changes during the callus culture cycle and differences between stock cultures. Can J Bot 53:261–271
Baque MA, Moh SH, Lee EJ, Zhong JJ, Paek KY (2012) Production of biomass and useful compounds from adventitious roots of high-value added medicinal plants using bioreactor. Biotechnol Adv 30(6):1255–1267
Benlarbi KH, Elmtili N, Macías FA, Galindo JCG (2014) Influence of *in vitro* growth conditions in the production of defence compounds in *Mentha pulegium* L. Phytochem Lett 8:233–244
Bertoli A, Lucchesini M, Mansuali-Sodi A, Leonardi M, Doveri S et al (2013) Aroma characterisation and UV elicitation of purple basil from different plant tissue cultures. Food Chem 141: 776–787
Berzin I, Cohen B, Mills D, Dinstein I, Merchuk JC (2000) RHIZOSCAN: a semiautomatic image processing system for characterization of the morphology and secondary metabolite concentration in hairy root cultures. Biotechnol Bioeng 70:17–24
Bondarev N, Reshetnyak O, Nosov A (2003) Effects of nutrient medium composition on development of *Stevia rebaudiana* shoots cultivated in the roller bioreactor and their production of steviol glycosides. Plant Sci 165:845–850
Bulgakov VP (2008) Functions of rol genes in plant secondary metabolism. Biotechnol Adv 26:318–324
Bulgakov VP, Shkryl YN, Veremeichik GN, Gorpenchenko TY, Inyushkina YV (2011) Application of *Agrobacterium* Rol Genes In: Alvarez MA (ed) Plant biotechnology: a natural phenomenon on secondary metabolism regulation. In: Genetic transformation. Croatia pp 262–270
Cai Z, Reidel H, Saw NMMT, Mewis I, Reineke K, Knorr D, Smetanska I (2011) Effects of elicitors and high hydrostatic pressure on secondary metabolism of *Vitis vinifera* suspension culture. Process Biochem 46:1411–1416
Carimi F, Zottini M, Formentin E, Terzi M, Schiavo FL (2003) Cytokinins: new apoptotic inducers in plants. Planta 216:413–421
Chaterjee C, Correll MJ, Weathers PJ, Wyslouzil BE, Walcerz DB (1997) Simplified acoustic window mist bioreactor. Biotechnol Tech 11(3):155–158
Condori J, Sivakumar G, Hubstenberger J, Dolan M, Sobolev V, Medina-Bolivar F (2010) Induced biosynthesis of resveratrol and the prenylated stilbenoids arachidin-1 and arachidin-3 in hairy root cultures of peanut: effects of culture medium and growth stage. Plant Physiol Biochem 48: 310–318
Costa P, Gonçalves S, Valentao P, Andrade PB, Romano A (2013) Accumulation of phenolic compounds in *in vitro* cultures and wild plants of *Lavandula viridis* L'Her and their antioxidant and anti-cholinesterase potential. Food Chem Toxicol 57:69–74
Curtin ME (1983) Harvesting profitable products from plant tissue culture. Biotechnology 1:649–657
Curtis WR (2005) Application of bioreactor design principles to plant micropropagation. Plant Cell Tissue Org Cult 81:255–264
Cusido RM, Onrubia M, Sabater-Jara AB, Moyano E, Bonfill M et al (2014) A rational approach to improving the biotechnological production of taxanes in plant cell cultures of Taxus spp. Biotechnol Adv. Available on line, doi: 10.1016/j.biotechadv.2014.03.002
Dalton CC, Street HE (1976) The role of the gas phase in the greening and growth of illuminated cell suspension cultures of spinach (*Spinacia oleracea* L.). In Vitro 12:485–494

de la Riva GA, González-Cabrera J, Vázquez-Padrón R, Ayra-Pardo C (1998) *Agrobacterium tumefaciens*: a natural tool for plant transformation. JB Electron J Biotechnol 1(3):118–131

Desai K, Badhe Y, Tambe SS, Kulkarni BD (2006) Soft-sensor development for fed-batch bioreactors using support vector regression. Biochem Eng J 27:225–239

Dewir YH, Chakrabarty D, Hahn EJ, Paek KY (2006) A simple method for mass propagation of *Spathiphyllum cannifolium* using airlift bioreactor. In Vitro Cell Dev Biol Plant 42:291–297

Dixon RA, Liu C, Jun JH (2013) Metabolic engineering of anthocyanins and condensed tannins in plants. Curr Opin Biotechnol 24:329–335

Donnez D, Kim KH, Antoine S, Conreux A, De Luca V et al (2011) Bioproduction of resveratrol and viniferins by an elicited grapevine cell culture in a 2 L stirred bioreactor. Process Biochem 46:1056–1062

Dörnenburg H (2004) Evaluation of immobilization effects on metabolic activities and productivity in plant cell processes. Process Biochem 39:1369–1375

Dörnenburg H, Knorr D (1995) Strategies secondary for the improvement of metabolite production in plant cell cultures. Enzym Microb Technol 17:674–684

Elwekeel A, Elfishway A, Abouzid S (2012) Enhanced accumulation of flavonolignans in *Silybum marianum* cultured roots by methyl jasmonate. Phytochem Lett 5:393–396

Ferri M, Dipalo SCF, Bagni N, Tassoni A (2011) Chitosan elicits mono-glucosylated stilbene production and release in fed-batch bioreactor cultures of grape cells. Food Chem 124:1473–1479

Flick CE, Evans DA, Sharp WR (1983) Organogenesis. In: Evans DA, Sharp WR, Ammirato PV, Yamada Y (eds) Handbook of plant cell culture. Macmillan, New York, pp 13–81

Floss DS, Schliemann W, Schmidt J, Strack D, Walter MH (2008) RNA interference mediated repression of MtCCD1 in mycorrhizal roots of *Medicago truncatula* causes accumulation of C27 apocarotenoids, shedding light on the functional role of CCD1. Plant Physiol 148: 1267–1282

Forcato DO, Pécora RP, Kivatinitz SC (2002) On-line biomass monitoring in bench-scale stirred bioreactors using parts of a liquid chromatography system. Biotechnol Lett 24:1999–2003

Fowler MW, Scragg AH (1988) Natural products from higher plants and plant cell culture. In: Pais MSS, Mavituna F, Novais JM (eds) Plant cell biotechnology, NATO ASI series. Springer, Berlin, pp 165–177

Fujita Y (1988) Industrial production of shikonin and berberine. Applications of plant cell and tissue culture. In: Ciba foundation symposium 137. Wiley, Chichester, pp 228–238

Gamborg OL, Miller RA, Ojima K (1968) Nutrient requirements of suspension cultures of soybean root cells. Expt Cell Res 50:151–158

Georgiev M, Pavlov A, Ilieva M (2006) Selection of high rosmarinic acid producing *Lavandula vera* MM cell lines. Process Biochem 41:2068–2071

Georgiev MI, Agostini E, Ludwig-Muller J, Xu J (2012) Genetically transformed roots: from plant disease to biotechnological resource. Trends Biotechnol 30(19):528–537

Giri A, Dhingra V, Giri CC, Singh A, Ward OP, Lakshmi Narasu M (2001) Biotransformations using plant cells, organ cultures and enzyme systems: current trend and future prospects. Biotechnol Adv 19:175–199

Goel MK, Mehrotra S, Kukreja AK, Shanker K, Khanuja SPS (2009) In vitro propagation of *Rauwolfia serpentina* using liquid medium, assessment of genetic fidelity of micropropagated plants, and simultaneous quantitation of reserpine, ajmaline, and ajmalicine. In: Protocols for *in vitro* cultures and secondary metabolite analysis of aromatic and medicinal plants. Methods in molecular biology. pp 17–33 © Humana Press, a part of Springer Science + Business Media, LLC 2009 doi: 10.1007/978-1-60327-287-2_2

Gonçalves S, Romano A (2013) *In vitro* culture of levanders (*Lavandula* spp.) and the production of secondary metabolites. Biotechnol Adv 31:166–174

Grant ME, Fuller KW (1971) Biochemical changes associated with the growth of root tips of *Vicia faba in vitro*, and the effect of 2,4-dichlorophenoxyacetic acid. J Exp Bot 22:49–59

Gueven A, Knorr D (2010) Isoflavonoid production by soy plant callus suspension culture. J Food Eng 103:237–243

Haberlandt G (1902) Kulturversuche mit isolierten Pflanzenzellen. Sitzungsber Akad Wiss Wien Math-Naturwiss Kl Abt J 111:69–92

Herrera-Estrella L, Depicker A, Van Montagu M, Schell J (1983) Expression of chimeric genes transferred into plant cells using a Ti-plasmid-derived vector. Nature 303(5914):209–213

Hu ZB, Du M (2006) Hairy root and its application in plant genetic engineering. J Integr Plant Biol 48(2):121–127

Huang TK, Mcdonald KA (2009) Bioreactor engineering for recombinant protein production in plant cell suspension cultures. Biochem Eng J 45:168–184

Huang C, Zhong JJ (2013) Elicitation of ginsenoside biosynthesis in cell cultures of *Panax ginseng* by vanadate. Process Biochem 48:1227–1234

Huang C, Qian ZG, Zhong JJ (2013a) Enhancement of ginsenoside biosynthesis in cell cultures of Panax ginseng by N′, N′-dicyclohexylcarbodiimide elicitation. J Biotechnol 165:30–36

Huang Y, Su CY, Kuo HJ, Chen YH, Huang PL, Lee KT (2013b) A comparison of strategies for multiple-gene co-transformation via hairy root induction. Appl Microbiol Biotechnol. doi:10.1007/s00253-013-5034-3

Hussain MS, Fareed S, Ansari S, Rahman MA, Ahmad IZ, Saeed M (2012) Current approaches toward production of secondary plant metabolites. J Pharm Bioallied Sci 4(1):10–20

Ionkova I, Momekov G, Proksch P (2010) Effects of cycloartane saponins from hairy roots of *Astragalus membranaceus* Bge., on human tumor cell targets. Fitoterapia 81:447–451

James E, Mills DR, Lee JM (2002) Increased production and recovery of secreted foreign proteins from plant cell cultures using an affinity chromatography bioreactor. Biochem Eng J 12: 205–213

Jung HJ, Kang SM, KANG YM, Kang MJ, Yun DJ, Bahk JD, Yang JK, Choi MS (2003) Enhanced production of scopolamine by bacterial elicitors in adventitious hairy root cultures of *Scopolia parviflora*. Enzym Microb Technol 33:987–990

Karwasara VS, Dixit VK (2012) Culture medium optimization for improved puerarin production by cell suspension cultures of *Pueraria tuberosa* (Roxb. ex Willd.) DC. In Vitro Cell Dev Biol Plant 48:189–199

Kevers C, Filali M, Petit-Paly G, Hagège D, Rideau M, Gaspar TH (1996) Habituation of plant cells does not mean insensitivity to plant growth regulators. In Vitro Cell Dev Biol Plant 32:204–209

Kim D, Pedersen H, Chin C (1990) Two stage cultures for the production of berberine in cell suspension cultures of *Thalictrum rugosum*. J Biotechnol 16:297

Kim J, Baek K, Son Y, Son S, Shin H (2009) Hairy root cultures of *Taxus cuspidata* for enhanced production of paclitaxel. J Appl Biol Chem 52:144–150

Kobayashi Y, Fukui H, Tabata M (1988) Berberine production by batch and semi-continuous cultures of immobilized *Thalictrum* cells in an improved bioreactor. Plant Cell Rep 7: 249–252

Kochkin DV, Kachala VV, Shashkov AS, Chizhov AO, Chirva VY, Nosov AM (2013) Malonyl-ginsenoside content of a cell-suspension culture of *Panax japonicus* var. repens. Phytochemistry 93:18–26

Komar S, Wittmann C, Heinzle E (2004) Minibioreactors. Biotechnol Lett 26:1–10

Lee SY, Kim YH, Roh YS, Myoung HJ, Lee KY, Kim DI (2004) Bioreactor operation for transgenic *Nicotiana tabacum* cell cultures and continuous production of recombinant human granulocyte-macrophage colony-stimulating factor by perfusion culture. Enzym Microb Technol 35:663–671

Lee EJ, Moh SH, Paek KY (2011) Influence of inoculum density and aeration volume on biomass and bioactive compound production in bulb-type bubble bioreactor cultures of *Eleutherococcus koreanum* Nakai. Bioresour Technol 102:7165–7170

Li W, Asada Y, Yoshikawa T (1998) Antimicrobial flavonoids from *Glycyrrhiza glabra* hairy root cultures. Planta Med 64:746–747

Lowry OH, Rosenbrough NJ, Farr AL, Randall RJ (1951) Protein measurement with the Folin phenol reagent. J Biol Chem 193:265

Lucchesini M, Bertoli A, Mensuali-Sodi A, Pistelli L (2009) Establishment of *in vitro* tissue cultures from *Echinacea angustifolia* D.C. adult plants for the production of phytochemical compounds. Sci Hortic 122:484–490

Luckner M (1990) Secondary metabolism in microorganisms, plants, and animals. Fischer-Verlag/Springer, Berlin, pp 15–21

Madhusudhan R, Ramachandra Rao S, Ravishankar GA (1995) Osmolarity as a measure of growth of plant cells in suspension cultures. Enzym Microb Technol 17:989–991

Malik S, Cusidó RM, Mirjalili HM, Moyano E, Palazón J, Bonfill M (2011) Production of the anticancer drug taxol in *Taxus baccata* suspension cultures: a review. Process Biochem 46:23–34

Mannan A, Shaheen N, Arshad W, Quershi RA, Zia M, Mirza B (2008) Hairy roots induction and artemisinin analysis in *Artemisia dubia* and *Artemisia indica*. Afr J Biotechnol 7:3288–3292

Marchev A, Haas C, Schulz S, Georgiev V, Steingroewer J, Bley T, Pavlov A (2013) Sage *in vitro* cultures: a promising tool for the production of bioactive terpenes and phenolic substances. Biotechnol Lett 36(2):211–221

Masakapalli SK, Ritala A, Dong L, van der Krol AR, Oksman-Caldentey KM, Ratcliffe RG, Sweetlove LJ (2014) Metabolic flux phenotype of tobacco hairy roots engineered for increased geraniol production. Phytochemistry 99:73–85

Medina-Bolivar F, Cramer C (2004) Production of recombinant proteins by hairy roots cultured in plastic sleeve bioreactors. In: Recombinant gene expression. Methods in molecular biology, Humana Press, Totowa, NJ, USA, vol 267. pp 351–363

Medina-Bolivar F, Condori J, Rimando AM et al (2007) Production and secretion of resveratrol in hairy root cultures of peanut. Phytochemistry 68:1992–2003

Mize CW, Chun YW (1988) Analyzing treatment means in plant tissue culture research. Plant Cell Tissue Org Cult 13:201–217

Mora-Pale M, Sanchez-Rodriguez SP, Linhardt RJ, Dordick JS, Koffas MAG (2013) Metabolic engineering and *in vitro* biosynthesis of phytochemicals and non-natural analogues. Plant Sci 210:10–24

Morini S, D'Onofrio C, Bellocchi G, Fisichella M (2000) Effect of 2,4-D and light quality on callus production and differentiation from *in vitro* cultured quince leaves. Plant Cell Tissue Org Cult 63:47–55

Mulder-Krieger TH, Verpoorte R, Baerheim Svendsen A, Scheffer JJC (1988) Production of essential oils and flavours in plant cell and tissue cultures. A review. Plant Cell Tissue Org Cult 13:85–154

Murashige T, Skoog F (1962) A revised medium for rapid growth and bio-assays with tobacco tissue cultures. Physiol Plant 15:473–479

Muthaiya MJ, Nagella P, Thiruvengadam M, Mandal AKA (2013) Enhancement of the productivity of tea (*Camellia sinensis*) secondary metabolites in cell suspension cultures using pathway inducers. J Crop Sci Biotechnol 16(2):143–149

Nadadoor VR, Siegler H, Shah SL, Mccaffrey WC, Ben-Zvi A (2012) Online sensor for monitoring a microalgal bioreactor system using support vector regression. Chemom Intell Lab Syst 110:38–48

Nagamori E, Hiroaka K, Honda H, Kobayashi T (2001) Enhancement of anthocyanin production form grape (*Vitis vinifera*) callus in a viscous additive-supplemented medium. Biochem Eng J 9:59–65

Nakagawa K, Konagai A, Fukui H, Tabata M (1984) Release and crystallization of berberine in liquid medium of *Thalictrum minus* cell suspension cultures. Plant Cell Rep 3:254

Nigra HM, Caso OH, Giulietti AM (1987) Production of solasodine by calli from different parts of *Solanum eleganifolium* Cav. plants. Plant Cell Rep 6(2):135–137

Nigra HM, Alvarez MA, Giulietti AM (1989) The influence of auxins, light and cell differentiation on solasodine production by *Solanum eleagnifolium* Cav. calli. Plant Cell Rep 8:230–233

Ono NN, Tian L (2011) The multiplicity of hairy root cultures: prolific possibilities. Plant Sci 180:439–446

Pacheco G, Garcia R, Lugato D, Vianna M, Mansur E (2012) Plant regeneration, callus induction and establishment of cell suspension cultures of *Passiflora alata* Curtis. Sci Hortic 144:42–47

Palacio L, Cantero JJ, Cusidó R, Goleniowski M (2011) Phenolic compound production by *Larrea divaricata* Cav. plant cell cultures and effect of precursor feeding. Process Biochem 46:418–422

Palacio L, Cantero JJ, Cusidó RM, Golienowski M (2012) Phenolic compound production in relation to differentiation in cell and tissue cultures of *Larrea divaricata* (Cav.). Plant Sci 193:1–7

Park JM, Giles KL, Songstad DD (1990) Production of sanguinarine by suspension culture of *Papaver somniferum* in bioreactors. J Ferment Bioeng 74:292–296

Park JM, Yoon SY, Giles KL, Songstad DD, Eppstein D et al (1992) Production of sanguinarine by suspension culture of *Papaver somniferum* in bioreactors. J Ferment Bioeng 74:292–296

Parsons J, Rodríguez Talou J, Giulietti AM, Alvarez María A (2001) Solasodine production by undifferentiated and differentiated *Solanum eleagnifolium* Cav. *in vitro* cultures. Phyton Int J Exp Bot 70:159–163

Patel H, Krishnamurthy R (2013) Elicitors in plant tissue culture. J Pharmacogn Phytochem 2(2):60–65

Pavlov A, Berkov S, Weber J, Bley T (2009) Hyoscyamine biosynthesis in *Datura stramonium* hairy root *in vitro* systems with different ploidy levels. Appl Biochem Biotechnol 157: 210–225

Peebles CAM, Sander GW, Hughes EH, Peacock R, Shanks JV, San KY (2011) The expression of 1-deoxy-D-xylulose synthase and geraniol-10-hydroxylase or anthranilate synthase increases terpenoid indole alkaloid accumulation in *Catharanthus roseus* hairy roots. Metab Eng 13(2):234–240

Pérez BA, Alvarez MA, Zanelli ML, Giulietti AM, Barreto DE (2001) Fungal extracts as elicitors of peroxidase production in root cultures of *Armoracia lapathifolia* transformed with *Agrobacterium rhizogenes*. Rev Investig Agropecuarias (RIA) 1:1–12

Prakash MG, Gurumurthi K (2010) Effects of type of explant and age, plant growth regulators and medium strength on somatic embryogenesis and plant regeneration in *Eucalyptus camaldulensis*. Plant Cell Tissue Organ Cult 100:13–20

Rai MJ, Shekhawat NS, Harish, Gupta AK, Phulwaria M, Ram K, Jaiswal U (2011) The role of abscisic acid in plant tissue culture: a review of recent progress. Plant Cell Tissue Organ Cult 106:179–190

Rajasekaran T, Ramakrishna A, Udaya Sankar K, Giridhar P, Ravishankar GA (2008) Analysis of predominant steviosides in *Stevia rebaudiana* bertoni by liquid chromatography/electrospray ionization-mass spectrometry. Food Biotechnol 22:179–188

Ramachandra Rao SR, Ravishankar GA (2002) Plant cell cultures: chemical factories of secondary metabolites. Biotechnol Adv 20:101–153

Randoux M, Jeauffre J, Thouroude T, Vasseur F, Hamama L et al (2012) Gibberellins regulate the transcription of the continuous flowering regulator, RoKSN, a rose TFL1 homologue. J Exp Bot 63(18):6543–6554

Rout GR, Samantaray S, Das P (2000) *In vitro* manipulation and propagation of medicinal plants. Biotechnol Adv 18:91–120

Russowski D, Maurmann N, Rech SB, Fett-Neto AG (2013) Improved production of bioactive valepotriates in whole-plant liquid cultures of *Valeriana glechomifolia*. Ind Crop Prod 46:253–257

Ruyter CM, Stöckgit J (1989) Neue Naturstoffe aus pflanzlichen Zell- und Gewebekulturen – eine Bestandsaufnahme. Git Fachz Lab 33:283–293

Ryu DDY, Lee SO, Romani RJ (1990) Determination of growth rate for plant cell cultures: comparative studies. Biotechnol Bioeng 35:305–311

Sabater-Jara AB, Pedreño MA (2013) Use of β-cyclodextrins to enhance phytosterol production in cell suspension cultures of carrot (*Daucus carota* L.). Plant Cell Tissue Organ Cult 114: 249–258

Sakuta M, Komamine A (1987) Cell growth and accumulation of secondary metabolites. In: Constabel F, Vasil YK (eds) Cell culture and somatic genetics of plants. Academic, San Diego/ New York, pp 97–114

Sánchez-Sampedro MA, Fernández-Tárrago J, Corchete P (2008) Some common signal transduction events are not necessary for the elicitor-induced accumulation of silymarin in cell cultures of *Silybum marianum*. J Plant Physiol 165:1466–1473

Savitha BC, Thimmaraju R, Bhagyalakshmi N, Ravishankar GA (2006) Different biotic and abiotic elicitors influence betalain production in hairy root cultures of *Beta vulgaris* in shake-flask and bioreactor. Process Biochem 41:50–56

Schenk RU, Hildebrandt AC (1972) Medium and techniques for induction and growth of monocotyledonous and dicotyledonous plant cell cultures. Can J Bot 50(1):199–204

Schübel H, Ruyter CM, Stöckigt J (1989) New phytochemicals from plant cell and tissue cultures. Bioeng News 10(16):2–3

Scragg AH, Kodja HP, Liu D, Merillon JM, Andreau F, Raideau M, Chenieux JC (1989) Stimulation par les cytokinins de I, accumulation d' alcaloides indoliques dans des suspensions cellularies de *Catharanthus roseus* (L.) G. Don Z Naturforsch 35c:551–556

Sharma M, Sharma A, Kumar A, Kumar Basu S (2011) Enhancement of secondary metabolites in cultured plant cells through stress stimulus. Am J Plant Physiol 6(2):50–71

Sharma P, Padh H, Shrivastava N (2013) Hairy root cultures: a suitable system for studying secondary metabolic pathways in plants. Eng Life Sci 1:62–75

Shinde AN, Malpathak N, Fulzele DP (2009) Studied enhancement strategies for phytoestrogens production in shake flasks by suspension culture of *Psoralea corylifolia*. Bioresour Technol 100:1833–1839

Shiota H, Kamada H (2008) Optical isomers of abscisic acid in carrot somatic embryos have the same effect on induction of dormancy and desiccation tolerance. Plant Biotechnol 25:457–463

Sivanandhan G, Vasudevan V, Theboral J, Selvaraj N, Ganapathi A, Manickavasagam M (2013) High frequency plant regeneration in *Withania somnifera* (l.) Dunal. Plant Cell Biotechnol Mol Biol 14(3 & 4):139–146

Skoric M, Todorovic S, Gligorijevic N, Jankovic R, Zivkovic S, Ristic M, Radulovic S (2012) Cytotoxic activity of ethanol extracts of *in vitro* grown *Cistus creticus* subsp. creticus L. on human cancer cell lines. Ind Crop Prod 38:153–159

Smolko DD, Peretti SW (1994) Stimulation of berberine secretion and growth in cell cultures of *Thalictrum minus*. Plant Cell Rep 14:131

Stäb MR, Ebel J (1987) Effects of $Ca2^{+}$ on phytoalexin induction by fungal elicitor in soybean cells. Arch Biochem Biophys 5157:416–423

Susila T, Satyanarayana Reddy G, Jyothsna D (2013) Standardization of protocol for in vitro propagation of an endangered medicinal plant *Rauwolfia serpentina* Benth. J Med Plants Res 7(29):2150–2153

Suzuki M, Nakagawa K, Fukui H, Tabata M (1988) Alkaloid production in cell suspension cultures of *Thalictrum flavum* and *T. dipterocarpum*. Plant Cell Rep 7:26

Tabata M, Fujita Y (1985) Production of shikonin by plant cell cultures. In: Zaitlin M, Day P, Hollaender A (eds) Biotechnology in plant science. Academic, Orlando, pp 207–218

Taha HS, El-Bahr MK, Seif-El-Nasr MM (2009) *In vitro* studies on Egyptian *Catharanthus roseus* (L.). II. Effect of biotic and abiotic stress on indole alkaloids production. J Appl Sci Res 5:1826–1831

Talano MA, Wevar Oller AL, González PS, Agostini E (2012) Hairy roots, their multiple applications and recent patents. Recent Pat Biotechnol 6:115–133. doi:10.2174/10115

Thanh NT, Murthy HN, Yu KW, Jeong CS, Hahn EJ, Paek KY (2006) Effect of oxygen supply on cell growth and saponin production in bioreactor cultures of *Panax ginseng*. J Plant Physiol 163:1337–1341

Thimmaraju R, Bhagayalakshmi N, Narayan MS, Ravishankar GA (2003) Kinetics of pigment release from hairy root cultures of *Beta vulgaris* under the influence of pH, sonication, temperature and oxygen stress. Process Biochem 38:1069–1076

Ulbrich B, Wiesner W, Arens H (1985) Large scale production of rosmarinic acid from plant cell cultures of *Coleus blumei* Benth. In: Deus-Neumann B, Barz W, Reinhard E (eds) Secondary metabolism of plant cell culture. Springer, Berlin, pp 293–303

Vahdati K, Bayat S, Ebrahimzadeh H, Jariteh M, Mirmasoumi M (2008) Effect of exogenous ABA on somatic embryo maturation and germination in Persian walnut (*Juglans regia* L.). Plant Cell Tissue Organ Cult 93:163–171

Verma AK, Singh RR, Singh S (2013) Improved alkaloid content in callus culture of *Catharanthus roseus*. Bot Serbica 36(2):123–130

Verpoorte R, Memelick J (2002) Engineering secondary metabolite production in plants. Curr Opin Biotechnol 13:181–187

Wallaart TE, Pras N, Beekman AC, Quax WJ (2000) Seasonal variation of artemisinin and its biosynthetic precursors in plants of *Artemisia annua* of different geographical origin: proof for the existence of chemotypes. Planta Med 66:57–62

Wheathers PJ, Bunk G, Mccoy MC (2005) The effect of phytohormones on growth and artemisinin production in *Artemisia annua* hairy roots. In Vitro Cell Dev Biol Plant 41:47–53

Wiktorowska E, Dlugosz M, Janiszowska W (2010) Significant enhancement of oleanolic acid accumulation by biotic elicitors in cell suspension cultures of *Calendula officinalis* L. Enzym Microb Technol 46:14–20

Wilson SA, Roberts SC (2014) Metabolic engineering approaches for production of biochemicals in food and medicinal plants. Curr Opin Biotechnol 26:174–182

Wink M (1988) Plant breeding: importance of plant secondary metabolites for protection against pathogens and herbivores. Theor Appl Genet 75:225–233

Wink M, Alfermann AW, Franke R, Wetterauer B, Distl M et al (2005) Sustainable bioproduction of phytochemicals by plant *in vitro* cultures: anticancer agents. Plant Genet Resour 3(2): 90–100

Wong PL, Royce AJ, Lee-Parsons CWT (2004) Improved ajmalicine production and recovery from *Catharanthus roseus* suspensions with increased product removal rates. Biochem Eng J 21:253–258

Wu J, Ge X (2004) Oxidative burst, jasmonic acid biosynthesis, and taxol production induced by low-energy ultrasound in *Taxus chinensis* cell suspension cultures. Biotechnol Bioeng 85:714–721

Wuang QM, Wuang L (2012) An evolutionary view of plant tissue culture: somaclonal variation and selection. Plant Cell Rep 31(9):1535–1547

Yamakawa T, Kato S, Ishida K, Kodama T, Minoda Y (1983) Production of anthocyanins by *Vitis* cells is suspension culture. Agric Biol Chem 47:2185–2191

Yeoman MM, Mitchell JP (1970) Changes accompanying the addition of 2, 4-D to excised Jerusalem artichoke tuber tissue. Ann Bot 34:799–810

Yesil-Celiktas O, Isleten M, Oyku Cetin E, Vardar-Sukan F (2010) Large scale cultivation of plant cell and tissue culture in bioreactors. Transworld Research Network, Kerala. pp 1–54

Yu KW, Murthy HN, Hahn EJ, Paek KY (2005) Ginsenoside production by hairy root cultures of *Panax ginseng*: influence of temperature and light quality. Biochem Eng J 23(1):53–56

Zenk MH, El-Shagi H, Arens H, Stockgit J, Weiler EW, Deus B (1977) Formation of the indole alkaloids serpentine and ajmalicine in cell suspension cultures. In: Barz W, Reinhard E, Zenk MH (eds) Plant tissue culture and its biotechnological application. Springer, Berlin, pp 27–43

Zhao R, Verpoorte R (2007) Manipulating indole alkaloid production by *Catharanthus roseus* cell cultures in bioreactors: from biochemical processing to metabolic engineering. Phytochem Rev 6:435–457

Zhao J, Davis L, Verpoorte R (2005) Elicitor signal transduction leading to production of plant secondary metabolites. Biotechnol Adv 23:283–333

Zhao ML, Shao JR, Tang YX (2009) Production and metabolic engineering of terpenoid indole alkaloids in cell cultures of the medicinal plant Catharanthus roseus (L.) G. Don (Madagascar periwinkle) Biotechnol Appl Biochem 52:313–323

Zobayed SMA, Murch SJ, Rupasinghe HPV, Saxena PK (2003) Elevated carbon supply altered hypericin and hyperforin contents of St. John's wort (*Hypericum perforatum*) grown in bioreactors. Plant Cell Tissue Organ Cult 75:143–149

Chapter 5
Solasodine Production in *Solanum eleagnifolium in vitro* Cultures

Abstract *Solanum eleagnifolium* Cav. var leprosum (Ortega) Dunal is a species that grow in Argentina. It is characterized for accumulating, mainly in green fruits, the glycoalkaloid solasodine, the nitrogenated analogue of diosgenine, which can be used for the synthesis of steroidal compounds (e.g.: contraceptives and corticosteroids).

In vitro cultures of *S. eleagnifolium* Cav. were established, and the conditions for achieving high solasodine yields were studied. Plant growth regulators balance, the age and size of the inocula, the conduction of the productive process in undifferentiated cultures and in hairy roots and transformed stem cultures were studied. Also, an approach to large-scale culture was performed. Plant growth balance appeared as the most significant strategy for improving solasodine yields in undifferentiated in vitro cultures of *S. eleagnifolium* Cav.

On the other hand, the solasodine levels attained in hairy root cultures were the highest, demonstrating that differentiation is related to the amount of for solasodine produced.

Keywords *Solanum eleagnifolium* Cav • Solasodine • *In vitro* cultures • Solasodine productivity • *Agrobacterium rhizogenes*

5.1 Introduction

Steroidal glycoalkaloids are a chemical group with particular biological activities that range from antimicrobial to analgesia, among others (Patel et al. 2013) (Table 5.1).

A representative of the group is solamargine (Fig. 5.1, II), a steroidal glycoalkaloid that can be used for the synthesis of steroidal compounds of pharmaceutical interest, such as steroidal contraceptives and corticosteroids (Mann 1978). Formerly, those compounds were produced exclusively from diosgenine, a steroidal glycoalkaloid present in the Dioscoreacea family. Their irrational exploitation caused a shortage of raw material with an increase of production costs promoting the search of alternative sources. Solasodine (Fig. 5.1, I) is the nitrogenated analogue of diosgenine, that can be used for the synthesis of 16-dehidropregnenolone and different steroidal hormones by chemical or microbial transformation (Rodriguez 1984). It is found in

M.A. Alvarez, *Plant Biotechnology for Health: From Secondary Metabolites to Molecular Farming*, DOI 10.1007/978-3-319-05771-2_5

Table 5.1 Some biological activities of steroidal glycoalkaloids found in species of the *Solanum* genus

Activity	Species	Active molecule/s	Reference
Analgesic	*S. xanthocarpum*	Solasodine	Gangwar et al. (2013)
Bronchodialator		Solasodine sapogenin	Kajaria et al. (2012)
Antiandrogenic			Khatodia et al. (2013)
Angiotensive	*S. laciniatum*	Solasodine	Shakirov et al. (2012)
Antineoplasic		Solasodine	Cham (2014)
		Rhamnosides	Cham (2008)
		Solamargine	
		Solasonine	
Anti-asmathic	*S. khasianum*	Solamargine	Tambe (2013)
Contraceptive		Solasonine	
Anticancerigen			

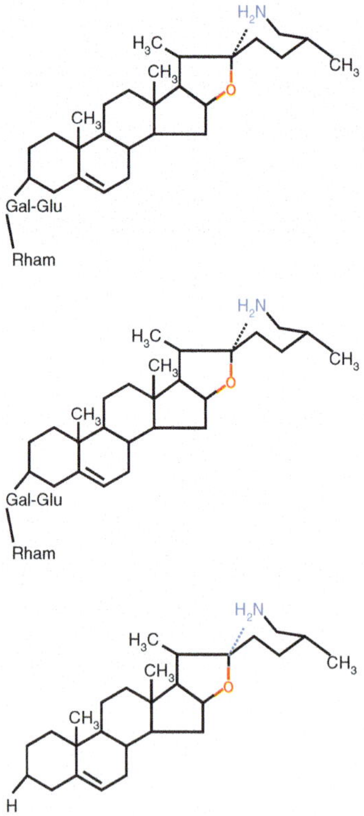

Fig. 5.1 Chemical structures of Solasonine (I), Solamargine (II) and Solasodine (III)

relatively high concentrations in leaves and fruits of a variety of species from the genus Solanum, as *S. laciniatum*, *S. aviculare*, *S. khasianum, S. marginatum*, and *S. eleagnifolium* (Rodríguez et al. 1979; Mann 1978; Guerreiro et al. 1971; Telek et al. 1977; Schreiber 1963; Schreuber 1968) (Table 5.2).

Table 5.2 Glycoalkaloid content of *Solanum* species from Argentina

Species	Plant organ	Glycoalkaloid content ($g\ kg^{-1}$ DW)	Glycoalkaloid
S. eleagnifolium Cav. var. leprosum	Green fruits	70–72	Solamargine
S. lorentzii Bit	Leaves and branches	10–37	Solamargine
S. calophyllum Phil	Leaves and branches	5.76	Solasodine
S. juncallense Reiche	Leaves and branches	5.05	Solamargine
S. sublobatum Willd	Fruits	11.60	Solamargine
S. atriplicifolium Gill	Fruits	–	–
S. euacanthum Phil	Leaves and branches	3.5	Solamargine
S. pyrethifolium Griseb	Leaves and branches	12.5	Solamargine
S. juvenale Thell	Leaves and branches	–	–

5.2 *Solanum eleagnifolim* Cav.

From the *Solanum* species that grow in Argentina, *S. eleagnifolium* Cav. var leprosum (Ortega) Dunal ("meloncillo del campo", "quillo", "revienta caballo") is the one with the highest solamargine content (Guerreiro et al. 1971). It is a perennial plant, 30–50 cm of height, with a branched stem covered by stellate trichomes and yellow stings. Leaves are numerous, alternated, covered by hairs and thorns along the petiole and nerves, both in the upper and lower surfaces. The radical system is dimorphic, with a vertical strong root, with diageotropically growth, that originates lateral roots that can extend to a distance of approximately 3 m. Flowers are blue, and are placed both in axial and terminal position. Fruits are berries of 6–8 mm diameter, each one with 40–50 seeds agglutinated by a saponin-rich substance. It is considered a toxic species especially for horses and cattle.

Usually, *S. eleagnifolium* Cav. grow in warm-temperate fertile soils, however, it can also be found in sandy or saline semiarid soils. In Argentina, it is distributed in the Cuyo region and in the Pampa cultivated areas.

Plants reproduce by seeds that germinate from September to December, in a first germination period, and in a second term in February. The presence of steroidal glycoalkaloids (e.g.: solamargine, solasonine) was established in different organs of the plants, but the highest levels are found in green fruits (Rodriguez 1984). In the plant, solasodine is present as a glycoside, whose sugars are bonded to the C3-hydroxyl group.

Several glycosides were described; the most frequent are solasonine and solamargine (Fig. 5.1, II), the last the most abundant. The glycosides are extracted from

Table 5.3 Methods for quantifying solasodine glycosides

Method	Sensibility	Reference
Gravimetric	Milligrams	Bell and Briggs (1942)
Tritimetric	Milligrams	Bite et al. (1970), Hardman and Williams (1971), Weiler et al. (1976)
Colorimetric	10–100 μg	Saini et al. (2007), Lancaster and Mann (1975)
Chromatographic	Nanograms	Trivedi and Pundarikakshudu (2007), Magrini et al. (1989), Crabbe and Fryer (1982), Herb et al. (1975)
Radioimmunoassay	Nanograms	Weiler et al. (1980)

S. eleagnifolium Cav. fruits by chemical or microbiological hydrolysis (Rodriguez 1984; Weston 1976; Regerat and Pourrat 1981; Ehmke 1986).

Solasodine can be determined by different analytical methods each one with different sensibility (Table 5.3): (a) Gravimetric (Bell and Briggs 1942), (b) Tritimetric (Bite et al. 1970), which detect milligrams of product, are fairly unspecific, and require an extensive purification that reduces the precision of the assays (Hardman and Williams 1971; Weiler et al. 1976), (c) Colorimetric, which is based on the development of colour by chemical reactions with the quaternary nitrogen of the molecule. The absorbance is read in a UV-visible spectrophotometer at a wavelength of 425 nm. The range of detection is 10–100 μg of solasodine (Saini et al. 2007; Lancaster and Mann 1975), (d) Chromatographic, as thin-layer chromatography (TLC), High Performance Thin Layer Chromatography (HPLC), and gas chromatography (GC) that are more sensitive than the formers (Trivedi and Pundarikakshudu 2007; Magrini et al. 1989; Crabbe and Fryer 1980; Herb et al. 1975), and (e) Radioimmunoassay which detects nanograms of solasodine glycosides (Weiler et al. 1980).

The production of solasodine was studied in plants growing in the field, focused on the analysis of the influence on yields of different factors such as influence of weather, nutrients, harvest method, processing of raw material, etc.

Nigra et al. (1990) have demonstrated the ability of *S. eleagnifolium* Cav. *in vitro* cultures to produce solasodine. The *in vitro* cultures are a recognized alternative production platform of secondary metabolites (Chap. 4). *In vitro* cultures have the advantage that it can be developed in controlled environmental conditions, free of pathogens, and with the possibility of a tight control of the phytofermentative process. With that background information, we have studied the *in vitro* establishment of *S. eleagnifolium* Cav. high producing lines through the selection and multiplication of solasodine-rich individuals or cultures. Also, we have studied the influence of plant growth regulators (PGR), and the establishment of transformed cultures (shoots and hairy roots) and their solasodine content.

The experimental scheme followed was: (a) Selection of producing and stable calli lines; (b) Study of the influence of PGR on the initiation and maintenance of clones and on cell growth and solasodine yield. These studies were performed both in calli as in cell suspension cultures; (c) Determination of the optimum inoculum size and age to initiate the phytofermentative process in suspension cultures; (d) Determination of the solasodine content in hairy roots and transformed stems obtained by transformation of *S. eleagnifolium* Cav. with *Agrobacterium* spp. (Fig. 5.2).

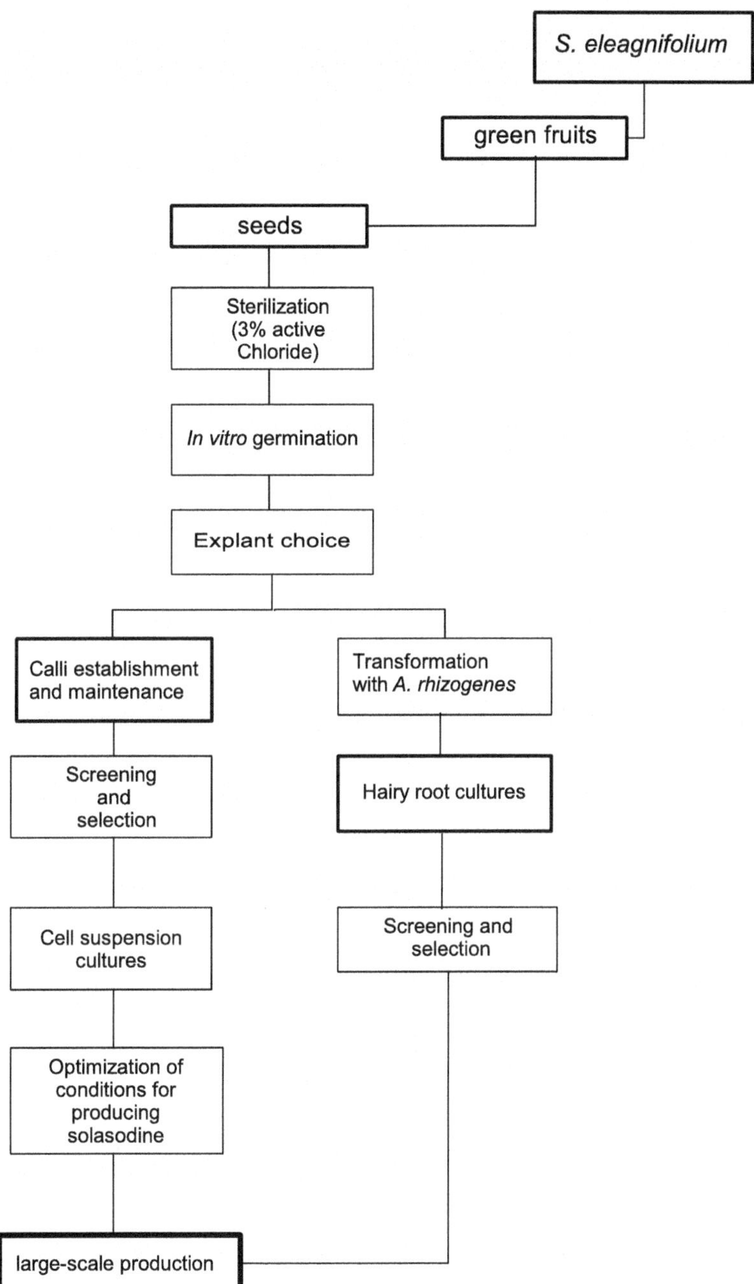

Fig. 5.2 General scheme of the procedure followed to analyze solasodine production by *S. eleagnifolium* Cav. *in vitro* cultures

5.3 Establishment of *S. eleagnifolium* Cav. *in vitro* Cultures

5.3.1 Establishment of Calli Culture

Calli were initiated according to Nigra et al. (1987). Briefly, seeds from green fruits collected in San Luis, Argentina, were washed and sterilized using NaClO (4 % active Cl2) and Tritón X-100 (0.1 %). Then, they were transferred to flasks containing MSRT culture medium (Nigra et al. 1987) with the addition of agar (8 g l^{-1}), and kept in growth chamber at 24±2 °C and a 16-hs photoperiod given by fluorescent lamps (irradiance 1.8 w m^{-1} seg^{-1}). After 3–4 weeks axenic plantlets developed from sterilized seeds (germination index: 40 %, contamination index: 5 %). Pieces of hypocotyls of approximately 1 cm length were transferred to flasks containing the same MSRT culture medium, with the addition of sucrose (30 g l^{-1}) and different auxins as PGR (Table 5.4). Flasks were maintained in a culture chamber at the same conditions described above.

Regions of hyperplasia appeared only in explants cultured in medium A after 8 days in culture. In the other media, morphological changes were only evident after 4 weeks in culture. After 8 weeks in culture, calli developed from explants in medium A, calli and thick roots grow in media B and C, the whole plant regenerated in medium D, and primordial non-proliferating buds developed in medium E.

Calli initiated in medium A were transferred to A, B, C, D, and E fresh media. The first weeks, calli proliferated in the four media but after 4 months calli turned necrotic in media C, D, and E, perhaps as a result of the disappearance of any residual 2,4-D content.

At the concentrations used, 2,4-D is the PGR of election for calli initiation, IBA does not induce calli, IAA induces the regeneration of the whole plant and NAA and 2,4,5-T resulted rhizogenic. Thus, IBA, IAA and NAA are not an adequate PGR for calli maintenance. Calli maintained in media A are green, friable, and sustains its growth rate (μ=0.08), doubling time (td=8.66 days), and growth index (15.5), and its viability and capability of producing solasodine for more than 2 years in *in vitro* conditions.

To evaluate solasodine content, samples are taken every 15 days (10 samples per each treatment), and maintained at 40 °C until constant weight. Solasodine was evaluated by the methods established by Lancaster and Mann (1975) and Magrini et al. (1989). The results were statistically analyzed by regression.

Solasodine content in calli varies according to the PGR used. Only 2,4-D and 2,4,5-T were able to maintain growth and solasodine production. In the case of 2,4,5-T, solasodine level is around an average value of 0.55±0.058 mg g DW^{-1}, which is similar to that reported in *S. laciniatum* Alt. In this case calli growing with 2,4-D (1–2 ppm) as PGR have a solasodine yield of 0.5 mg g DW^{-1} (Hosoda et al. 1979). On the other hand, 2,4-D as PGR gave a highest solasodine initial amount (1.3 mg g DW^{-1}) during the first subcultures but the solasodine yield fell down in the subsequent subcultures (Fig. 5.3). In conclusion, 2,4,5-T allows constant relatively high solasodine amounts while 2,4-D allows highest solasodine amounts during the first months in culture but the productive capacity of the cultures decreases with time in culture (Nigra et al. 1989).

Calli lines were identified and selected according to their solasodine yield. They were classified in low producing lines (<1 mg solasodine g DW^{-1}), lines that initially

Table 5.4 Culture media used to establish *S. eleagnifolium* Cav. *in vitro* cultures

Medium	Basal medium	Auxins
A	MSRT	2,4-D (4.5 μM)
B	MSRT	2,4,5-T (3.91 μM)
C	MSRT	NAA (5.37 μM)
D	MSRT	IAA (5.70 μM)
E	MSRT	IBA (4.90 μM)

2,4-D: 2,4- dichlorophenoxy acid; 2,4,5-T: 2,4,5-trichlorophenoxy acid; NAA: naphthalene acid; IAA: indole acetic acid; IBA: indole butyric acid

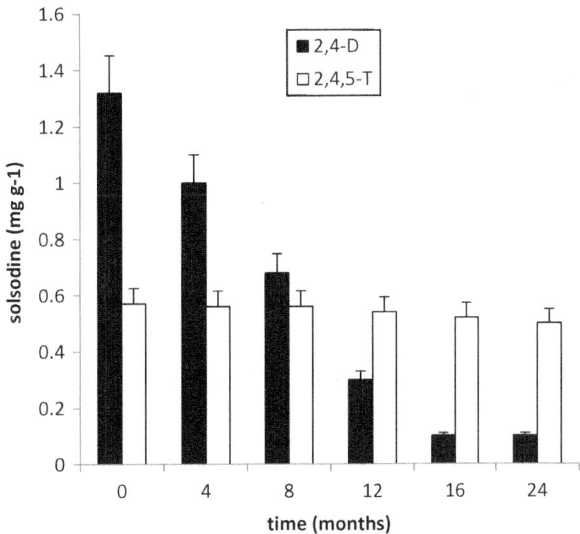

Fig. 5.3 Variation with time of solasodine content in *S. eleagnifolium* Cav. calli maintained in medium MSRT supplemented with 2,4-D (4.5 μM) or 2,4,5-T (3.91 μM) as PGR

showed high solasodine content (>4 mg solasodine g DW^{-1}) that fell down after several subcultures, and lines with a stable solasodine content (around 1.5 mg solasodine g DW^{-1}). From the lines developed, line 009 was selected for further experiments considering its high and stable solasodine yield.

5.3.2 Establishment of Cell Suspension Cultures

Friable calli from the selected line were transferred to 200 ml-Erlenmeyer flasks containing 40 ml MSRT culture medium without agar and with the addition of different PGR (Table 5.5). Cultures were incubated at 24 ± 2 °C, 16-h photoperiod (irradiance = 1.8 w m^{-2} s^{-1}) in rotary shaker at 120 rpm. The obtained cell suspensions were transferred to identical fresh culture medium each 15th day, samples were taken in each subculture to analyze the cells microscopically.

Cell suspensions initiated in media C and D were heterogeneous, with green-brownish clumps, that show free cells or in small aggregates under the optical microscope. In media A and B cell suspensions are homogeneous, showing tubular

Table 5.5 Culture media used to initiate cell suspension cultures of *S. eleagnifolium* Cav

Culture medium	Basal medium	Auxins
A	MSRT	2,4-D (4.5 μM)
B	MSRT	2,4,5-T (3.91 μM)
C	MSRT	NAA (5.37 μM)
D	MSRT	IAA (5.70 μM)

free cells with thin walls under the optical microscope. The presence of aggregates in medium C and D do not allow a homogeneous transference of mass and energy as happen in homogeneous cell suspension cultures. Thus, cell suspension cultures in media A and B were selected for further experiments since the homogeneity of transference is a key factor for conducting a biotechnological process.

5.4 Influence of Plant Growth Regulators

In order to analyze the influence of PGR on cell growth and solasodine production, 7-days old suspended cells maintained in medium A were filtered through sterile plastic sieves (40 mm mesh diameter) to retain cell aggregates, and transferred to 40 ml Erlenmeyer flasks, at a 5 % V/V inoculum size, containing one of the media described in Table 5.6. Cultures were developed at the same conditions described above during 25 days, every 5 days 3 ml-samples were harvested to determine packed cell volume (PCV), dry weight (DW) and solasodine content. Growth Index and solasodine productivity are shown in Table 5.7.

Cell viability was determined combining the Evans blue method with a treatment with fluorescein diacetate (FDA). Viable cells do not embody the blue dye and are fluorescent under the UV-light. Viability index (VI) was determined by determining the percent relationship among live cells and dead cells (Alvarez 1992, PhD Thesis).

Figure 5.4 shows the cell growth and viability in cell suspension cultures growing in medium A. The growth curve shows no initial lag phase, an exponential phase with a constant growth up to the 10th day of culture, followed by a stationary phase that maintains up to the end of the culture with a final biomass of 4.2 mg DW ml^{-1}. Under the microscope cells are isodiametric in the early exponential phase, and then they elongate and bond by their minor walls in aggregates of no more than 4–6 cells. As for the viability index, it stays over 80 % during the first 25 days, being the optimal timing for transference between the 10th and 20th day (VI=98 % and 85 % respectively) (Fig. 5.4).

To analyze the effect of different relationships auxin: citoquinine on solasodine production, cell suspension cultures were transferred to MSRT medium with each one of the PGR relationship showed in Table 5.8. Cultures were developed during 25 days, 3 ml-samples were taken every 5 days to analytical studies.

After 10 days in culture, the highest GI and solasodine content were attained with low concentrations of kinetin (0.25, 0.50 and 1.0 μM). The productivity reached in the presence of NAA (2.54±0.22 mg l^{-1} day^{-1}) was significant higher ($P<0.05$) to those attained in the presence of 2,4-D (0.50 mg l^{-1} day^{-1}) (Table 5.9). The growth curves

Table 5.6 Culture media used to analyze the influence of PGR on cell growth and solasodine content in *S. eleagnifolium* Cav. cell cultures

Culture medium	Basal medium	Auxins (μM)	
A1	MSRT	2,4-D	0.5
A2	MSRT		5
A3	MSRT		50
B1	MSRT	NAA	0.5
B2	MSRT		5
B3	MSRT		50
C1	MSRT	IAA	0.5
C2	MSRT		5
C3	MSRT		50
D1	MSRT	2,4,5-T	0.5
D2	MSRT		5
D3	MSRT		50

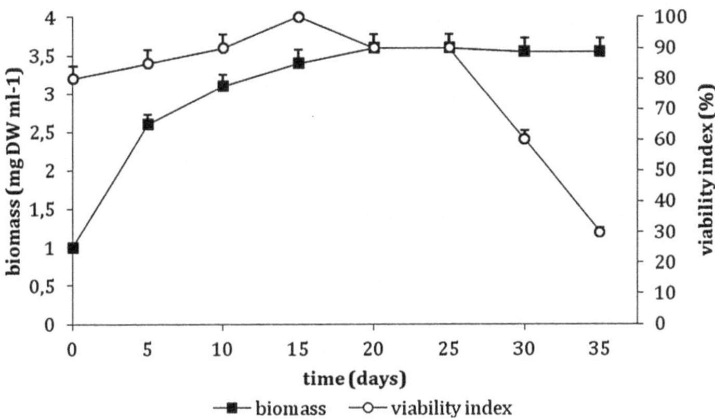

Fig. 5.4 Cell growth (mg DW biomass ml^{-1}) and viability index (%) of *S. eleagnifolium* Cav. cell suspension cultures in MSRT media with 2,4-D 4.5 μM

show that the higher productivity corresponds to the 10th day of culture and immediately sharp fall down, probably due to a solasodine degradation process (Fig. 5.4).

In suspension cell cultures growing in MSRT media with 2,4-D as PGR the specific growth rate is high ($\mu = 0.34$ days^{-1}) but the solasodine productivity is poor (0.58 ± 0.15 mg l^{-1} day^{-1}) while in the presence of NAA (50 μM) and kinetin (0.25 μM) as PGR, cells grow slowly ($\mu = 0.14$ days^{-1}) but solasodine productivity and yield are higher (2.54 ± 0.22 mg l^{-1} day^{-1} and 6.98 mg g^{-1} respectively) (Fig. 5.5). The production of solasodine resulted strongly influenced by the PGR relationship; it is inhibited by a high 2,4 D concentration and by the PGR 2,4,5-T and IAA. From the auxins tested, NAA seems to be the most effective promoting solasodine production particularly in the presence of kinetin (0.25 μM) (Alvarez et al. 1993). These results are in according to those obtained for the production of anthraquinone by *Morinda citrifolia* and *Rubia cordifolia* (Zenk et al. 1975; Suzuki et al. 1987),

Table 5.7 Variation of growth index (GI) and productivity (mg gDW^{-1} l^{-1}) os *S. eleagnifolium* cell suspension cultures after 10 days in culture in the different media tested

Auxin (μM)		Growth index (%)	Solasodine productivity (mg l^{-1} day^{-1})
2,4-D	0.5	14	0.60
	5	13	0.60
	50	12	0.50
2,4,5-T	0.5	1	0.05
	5	0.8	0.20
	50	0.2	0.28
IAA	0.5	3	0.22
	5	5	0.40
	50	3	0.30
NAA	0.5	19	0.30
	5	15	0.48
	50	19	0.80

ANOVA analysis was performed in all the experiments. From all the PGR tested, 2,4-D and NAA have a positive effect both on cell growth and solasodine production

Table 5.8 Different hormone relationships used to evaluate solasodine productivity of *S. eleagnifolium* Cav. cell suspension cultures

Kinetin (μM)	2,4-D (μM)	2,4,5-T (μM)	NAA (μM)	IAA (μM)
0	0.5	0.5	0.5	0.5
	5	5	5	5
	50	50	50	50
0.25	0.5	0.5	0.5	0.5
	5	5	5	5
	50	50	50	50
0.5	0.5	0.5	0.5	0.5
	5	5	5	5
	50	50	50	50
5.0	0.5	0.5	0.5	0.5
	5	5	5	5
	50	50	50	50
10	0.5	0.5	0.5	0.5
	5	5	5	5
	50	50	50	50

Table 5.9 Influence of kinetin on growth index (GI) and productivity of *S. eleagnifolium* Cav. cell suspension cultures

Kinetin (μM)	Auxin (μM)		Growth index (%)	Solasodine productivity (mg l^{-1} day^{-1})
0 (control)	2.4-D	0.5	14.8	0.58
0.25		0.5	2.0	1.0
0.5		0.5	1.9	1.3
1.0		0.5	2.8	1.4
0.25	NAA	50	4.0	2.5
0.5		50	0.6	1.9
1.0		50	0.4	1.8

C: control MSRT medium with 2,4-D (0.5 μM). Culture period: 10 days

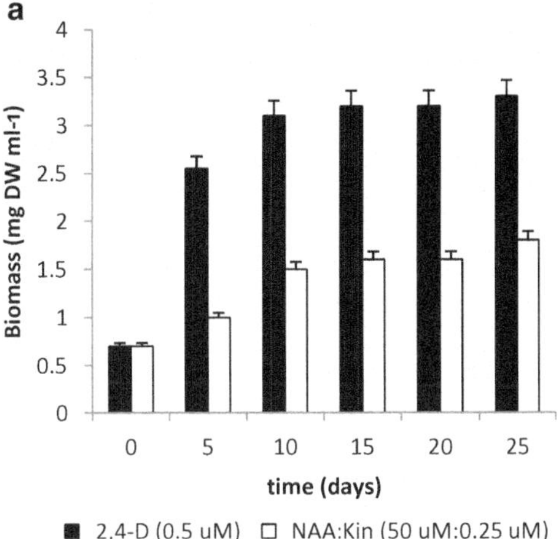

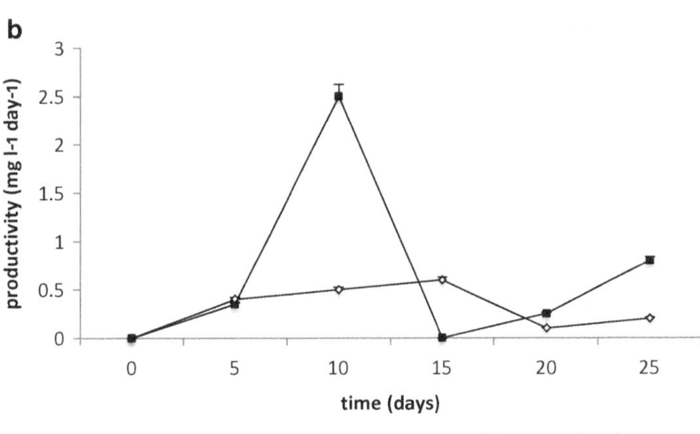

Fig. 5.5 (**a**) Cell growth and (**b**) solasodine productivity (at the 10th day of culture) in cell suspension cultures of *S. eleagnifolium* growing in MSRT+2,4-D (0.5 μM) or MSRT+NAA (50 μM): kinetin (0.25 μM)

rosmarinic acid by *A. officinalis* (De-Eknamul and Ellis 1985), and berberine by *T. minus* (Nakagawa et al. 1986).

The experiments also show that the PGR relationship for optimal growth is different from those for optimal solasodine production, which suggest that a two-step productive process could be successful, as it was for the production of berberine by *Berberis buxifolia* Lam. (Alvarez et al. 2009). In such a dual system *S. eleagnifolium* Cav. cells could proliferate in MSRT medium with 0.5 μM 2,4 D as PGR and then they could be transferred to a productive medium with NAA (50 μM) and Kinetin (0.25 μM) as PGR.

The combination of other citoquinines (BAP and zeatin) with auxins (2,4-D 0.5 μM and NAA 50 μM) was tested for studying their effect on solasodine production (Table 5.10). Cell growth increased at all the combinations tested, which is in agreement with the fact that the external supply of citoquinines (especially of a

Table 5.10 Influence of zeatin (Z) and 6-benzyl aminopurine (BAP) on growth index (GI) and solasodine yield in *S. eleagnifolium* Cav. cell suspension cultures

Citoquinine (μM)		Auxin (μM)		GI	Solasodine yield (mg g^{-1})
Zeatin	0	NAA	50	12.5	3.2
	0.25			40.5	0.8
	0.50			31.0	3.8
	1.0			31.0	1.2
BAP	0.25		50	25	2.9
	0.50			29	1.0
	1.0			11	2.0
Zeatin	0.25	2,4-D	0.50	30	5.0
	0.50			39	1.1
	1.0			40	2.0
BAP	0.25		0.50	15	4.8
	0.50			17	6
	1.0			15	2.8

C: control medium (MSRT medium + 2,4-D 0.5 μM). Culture period: 20 days

natural citoquinine as zeatin) induces cell division (Nishinari and Syono 1980). As for solasodine yield, with the relationship 2,4-D (0.5 μM): BAP (0.5 μM) a value of 6.00 ± 0.04 mg g^{-1} was attained that is comparable to those obtained with NAA (50 μM): kinetin (0.25 μM). For all the PGR relationship tested, solasodine was not detected in the culture medium.

As a conclusion, from the assays performed for determining the influence of PGR on solasodine production by *S. eleagnifolium* Cav. the best amounts were obtained with NAA: Kinetin (0.5 μM:0.25 μM) and 2,4-D: BAP (0.5 μM:0.5 μM) which are comparable to those measured in the wild plant (Guerreiro et al. 1971).

5.5 Influence of Inoculum Age and Size

For inoculum size experiments, 21-days-old cell suspension maintained in MSRT medium with 2,4-D (5 μM) were washed five times with MSRT medium without PGR and then transferred to 200 ml-Erlenmyer flasks containing 40 ml MSRT medium with NAA: kinetin (50 μM: 0.25 μM) as PGR. The inoculum sizes tested were 5 %, 10 % and 20 % v/v.

For inoculum age experiments, 3, 8, 15, and 19-days-old-cell suspension cultures at a 5 % inoculum size were inoculated in 200 ml Erlenmyer flasks containing 40 ml MSRT culture medium with 2,4-D: kinetin (5 μM: 0.25 μM) as PGR. Cultures were maintained during 25 days at 24 ± 2 °C, 16-h photoperiod in a rotary shaker at 120 rpm. Samples of 3 ml were taken each 5 days for determining biomass and solasodine content.

From the inoculum sizes tested, the highest biomass (3.6 mg g DW^{-1}) was attained at a 20 % v/v inoculum size ($P<0.10$) and the highest solasodine yield with an inoculum size of 5 % (4.0 ± 0.35 mg g DW^{-1}). Fujita et al. (1985) working with *L. erythrorhizon* for producing shikonine also observed that converse relationship between yields and inoculum size. As for the inoculum age, a 19-days old inoculum produces the highest solasodine yields (2.7 ± 1.51 mg g DW^{-1}) but no significant differences were observed in biomass yield ($P<0.05$) (Parsons et al. 2001).

5.6 Transformed Stems and Hairy Roots

Previous studies (Nigra et al. 1989) had demonstrated that solasodine yields increased in the presence of certain degree of differentiation induced by PGR relationship.

On the other hand, Lancaster and Mann (1975) reported that for *S. laciniatum* Ait, the solasodine concentration in the wild plant reached their highest value in green fruits but that during fruit maturation the alkaloid disappeared. Czygan and Willuhn (1967) suggested that the alkaloid metabolites could be involved in carotenoid biosynthesis in the mature fruits. Also, Lancaster and Mann (1975) demonstrated that the decrease in the total solasodine amounts were accompanied by an increase in the total amount in roots, suggesting that solasodine cold be transported from the mature organs to the roots to accumulate there. That phenomenon of translocation of metabolites is a quite general phenomenon described in several species. Based on those observations, the analysis of solasodine production in transformed organs was performed considering their characteristics of higher genetic and productive stability.

5.6.1 Hairy Root Cultures

A. rhizogenes LBA strains 9402 and 9816 were maintained in YMB medium. Stems and leaves of *S. eleagnifolium* Cav. axenic plantlets were infected with 48-h *A. rhizogenes* cultures cultured at 28 °C, with the addition of acetosyringone (0.1 % P/V). Infected explants were transferred to MSRT solid medium without the addition of PGR and cultured at 24 ± 2 °C and 16-h photoperiod. After 2–3 weeks root apex that appeared in the infection sites, were excised and transferred to MSRT culture medium in the presence of ampicillin 0.2 % to eliminate *Agrobacterium*.

When the cultures showed absence of *Agrobacterium*, the roots were transferred to MSRT medium without the addition of PGR and antibiotics. Transformation was confirmed by PCR using primers for *rolB* to amplify a 700 bp fragment as was described by Hamill et al. (1989). The transformation efficiency was 80 %.

Several clones were obtained, with different morphological characteristics and fast growth. The clones 1 and 2 were 2 the best productive with solasodine yields of 1.9 ± 0.3 mg g^{-1} and 1.175 ± 0.3 mg g^{-1} respectively.

5.6.1.1 Kinetic of Growth and Production of Solasodine of *S. eleagnifolium* Cav. Hairy Roots

Clone 1 was used for performing these experiments for their characteristics of growth and solasodine production. Apexes of hairy roots (3.5 mg ± 0.5 mg DW) were transferred to 250 ml Erlenmeyer flasks containing 40 ml MSRT medium without PGR. Cultures were performed at 24 ± 2 °C, 16-h photoperiod, 1.8 w m^{-2} s^{-1} irradiance and 100 rpm during 30 days. Samples of 3 Erlenmeyers were taken each 5 days to evaluate growth and solasodine content.

The kinetic of growth and solasodine production is shown in Fig. 5.6. In hairy roots a lag phase is followed by an exponential phase of 10 days before entering in

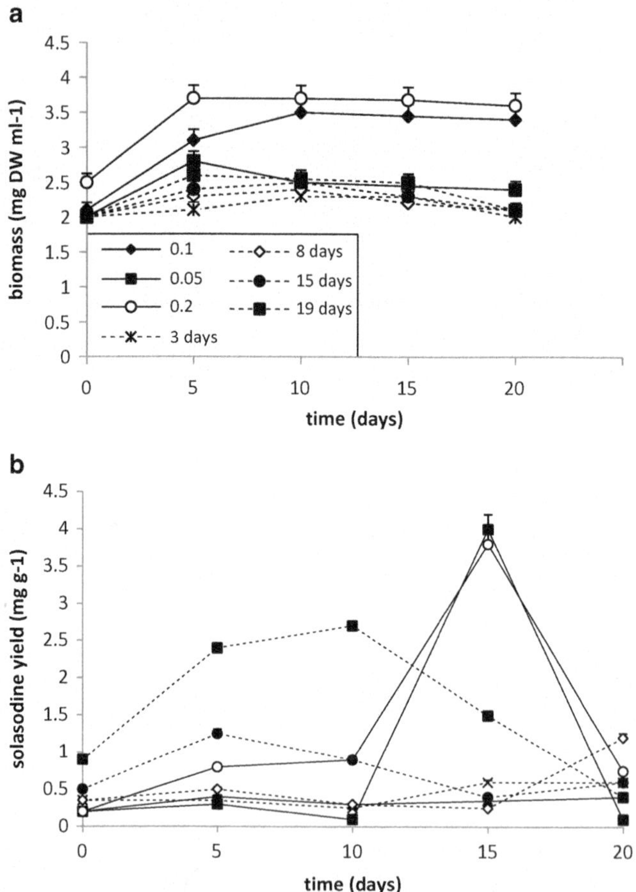

Fig. 5.6 Influence of inoculum size and age on biomass (**a**) and solasodine (**b**) yields by *S. eleagnifolium* Cav. cell suspension cultures in MSRT culture medium with NAA (50 μM): Kinetin (0.25 μM) as PGR

Table 5.11 Comparison of the specific growth rate, growth index, doubling time, final biomass, and solasodine yield of hairy roots and calli growing in MSRT medium with 2,4-D (4.5 μM) and 2,4,5-T (3.91 μM)

Parameter	Hairy roots	Calli MSRT + 2,4-D (4.5 μM)	Calli MSRT + 2,4,5-T (3.91 μM)	Cell suspension cultures MSRT + NAA: kin (50 μM: 0.25 μM)	Cell suspension cultures MSRT + 2,4-D: BAP (0.5 μM: 0.5 μM)
μ (days)	0.55	0.08	0.08	0.14	0.54
td	30 h	8.66 days	8.45 days	4.95	2.03
Final biomass (mg FW^{-1})	140	180	104	83	102
GI	39	45	41	4	17
Solasodine yield (mg g^{-1})	1.90 ± 0.08	1.3 ± 0.08	0.55 ± 0.09	6.98 ± 0.5	6.00 ± 0.5

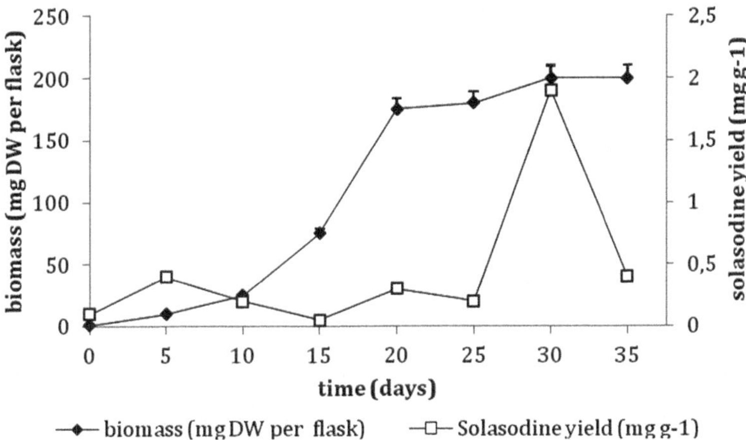

Fig. 5.7 Kinetic of growth and solasodine yield of *S. eleagnifolium* Cav. hairy root, clone 1

stationary state (Alvarez et al. 1994). A comparison among the specific rate of growth, growth index, doubling time, final biomass and solasodine yield of hairy roots and calli rowing gin MSRT medium with 2,4-D (4.5 μM) and 2,4,5-T (3.91 ± M) are shown in Table 5.11 (Fig. 5.7).

5.6.2 Transformed Stems

Plantlets were transformed using 48 h old-culture of *A. tumefaciens* strain T37 maintained in trypton-yeast culture medium (Ty) at 28 °C. Infected plantlets were transferred to solid MSRT culture medium at 24±2 °C, 16 h photoperiod. When the buds appeared at the infection points, they are transferred to MSRT liquid medium containing ampicillin (0.5 %) to eliminate the bacteria. Transformation was confirmed according to Otten and Schilperoot (1978). The efficiency of transformation was 76 %. Several clones were obtained; the independence of the exogenous supply of PGR is evident, as was also determined by Spencer et al. 1990; Gresshoff et al. 1979; Ooms et al. 1981. We selected those that produced significant amounts of solasodine. In average, the growth index was 54.4 and the solasodine yield was 0.35±0.09 mg g^{-1} which is comparable to the values obtained in stems and leaves of the wild plant and was lower to those values in the best performing undifferentiated *in vitro* cultures.

5.7 Conclusive Remarks

Different strategies were used to analyze solasodine production by *S. eleagnifolium* Cav. *in vitro* cultures. Plant growth balance appeared as the most significant strategy for improving solasodine yields in undifferentiated *in vitro* cultures of *S. eleagnifolium* Cav. The obtained values with some PGR relationships are comparable to those of green fruits of the wild plant.

Through screening and selection it was possible to obtain strains with different productivity along time being the yields constants when using 2,4,5-T as PGR or with high initial values that decreased with time when using 2,4-D as PGR. The cultures of suspended cells were homogeneous, mostly of single cells or small aggregates when using 2,4,5-T for initiation of cultures. The highest yields were attained with NAA (50 μM): kinetin (0.25 μM) and 2,4-D (0.5 μM): BAP (0.5 μM), which were higher than those determined in stems and leaves of the wild type plant. The fact that differentiation is related to the amount of secondary metabolites produced was confirmed after establishing hairy root and transformed stem cultures obtained by infection of *S. eleagnifolium* Cav. plantlets with *A. rhizogenes*. Both transformed cultures produced a solasodine yield that was also higher than those of the aerial parts of the wild type plant.

References

Alvarez MA (1992) Influencia de reguladores de crecimiento y organogenesis sobre la producción de solasodina por cultivo *in vitro* de *Solanum eleagnifolium* Cav. PhD thesis, Facultad de Farmacia y Bioquímica, Universidad de Buenos Aires

Alvarez MA, Nigra HM, Giulietti AM (1993) Solasodine production by *Solanum eleagnifolium* Cav. *in vitro* cultures: influence of plant growth regulators, age and inoculum size. Large-scale production. Nat Prod Lett 3(1):9–19

Alvarez MA, Rodríguez Talou J, Paniego N, Giulietti AM (1994) Solasodine Production in transformed cultures (roots and shoots) of *Solanum eleagnifolium* Cav. Biotechnol Lett 16(4):393–396

Alvarez MA, Fernandez Eraso N, Pitta-Alvarez S, Marconi PL (2009) Two-stage culture for producing berberine by cell suspension and shoot cultures of *Berberis buxifolia* Lam. Biotechnol Lett 31:457–463

Bell RC, Briggs LH (1942) *Solanum* alkaloids. Part I. The alkaloid from the fruit of *Solanum aviculare*. J Chem Soc 1:1–2

Bite P, Magó-Karácsony E, Rettegi T, Uskert A (1970) *Solanum* Glycosides V. Acta Chim Hungarica: J Hungarian Acad Sci 64(2):199–201

Cham BE (2008) Cancer intralesion chemotherapy with solasodine rhamnosyl glycosides. Res J Biol Sci 3(9):1008–1017

Cham BE (2014) A review of solasodine rhamnosides therapy for *in-situ* squamous cell carcinoma on the penis. Br J Med Med Res 4(2):621–631

Crabbe PG, Fryer C (1980) Rapid quantitative analysis of solasodine, solasodine glycosides and solasodine by high-pressure liquid chromatography. J Chromatogr 187(1):87–100

Crabbe PG, Fryer C (1982) Evaluation of chemical analysis for the determination of solasodine in Solanum laciniatum. J Pharm Sci 71(12):1356–1362

Czygan FC, Willuhn G (1967) Untersuchungen Über Gehaltsänderungen der Lipochrome und Steroidalkaloide in Rot– und Gelbfrüchtigen Tomaten (*Lycopersicon esculentum* L.) Verschiedener Reifegrade. Planta Med 15:404–415

De-Eknamul W, Ellis BE (1985) Effects of auxins and cytokinins on growth and rosmarinic acid formation in cell suspension cultures of *Anchusa officinalis*. Plant Cell Rep 4:50–53

Ehmke A (1986) An inducible saponine cleaving glycosidase from fusarium and its application to the production of solasodine. Planta Med 52(6):507

Fujita Y, Takahashi S, Yamada Y (1985) Selection of cell lines with high productivity of shikonin derivatives by protoplast culture of *Lithospermum erythrorhizon* cells. Agric Biol Chem 49(6): 1755–1759

Gangwar AK, Ghosh AK, Saxena V (2013) Phytochemical screening and analgesic activity of "Kantkari". Int J Herb Med 1(2):177–186

Gresshoff PM, Skotnicki ML, Rolfe BG (1979) Crown gall teratoma formation is plasmid and plant controlled. J Bacteriol 137:1020–1021

Guerreiro E, Giordano OS, Kavka J, D'Arcangelo AT (1971) Aislamiento de compuestos esteroidales de *Solanum eleagnifolium* Cav. Anales Química 67:789–794

Hamill JD, Robins RJ, Rhodes MJC (1989) Alkaloid production by transformed root cultures of *Cinchona ledgeriana*. Planta Med 55:354–357

Hardman R, Williams TG (1971) The titrimetric assay of solasodine, a spirosolane of commercial importance from *Solanum laciniatum* Ait. J Pharm Pharmacol 23(S1):231S

Herb SF, Fitzpatrick TJ, Osman SF (1975) Separation of potato glycoalkaloids by gas-chromatography. J Agric Food Chem 23:520

Hosoda N, Ito H, Yatazawa M (1979) Some accounts on cultural conditions of callus tissues of *Solanum laciniatum* Ait. for producing solasodine. Agric Biol Chem 43(8):1745–1748

Kajaria DK, Gangwar M, Kumar D, Sharma AK, Tilak R et al (2012) Evaluation of antimicrobial activity and bronchodilator effect of a polyherbal drug-Shrishadi. Asian Pac J Trop Biomed 2(11):905–909

Khatodia S, Biswas K, Bhatotia K (2013) Induction and establishment of hairy root culture of *Solanum xanthocarpum* using *Agrobacterium rhizogenes*. J Pharm BioSci 1:59–63

Lancaster JE, Mann JD (1975) Changes in solasodine content during the development of *Solanum laciniatum* Air. NZ J Agric Res 18:139–144

Magrini E, Giulietti AM, Wilson E, Cascone O (1989) High performance liquid chromatographic determination of glycoalkaloids in callus and fruits of *Solanum eleagnifolium*. Biotechnol Tech 3:185–188

Mann JD (1978) Production of solasodine for the pharmaceutical industry. Adv Agron 30: 207–243

Nakagawa K, Fukui H, Tabata M (1986) Hormonal regulation of berberine production in cell suspension cultures of *Thalictrum minus*. Plant Cell Rep 5(1):69–71

Nigra HM, Caso OH, Giulietti AM (1987) Production of solasodine by calli from different parts of *Solanum eleagnifolium* Cav. plants. Plant Cell Rep 6:135–137

Nigra HM, Alvarez MA, Giulietti AM (1989) The influence of auxins, light and cell differentiation on solasodine production by *Solanum eleagnifolium* Cav. Plant Cell Rep 8:230–233

Nigra HM, Alvarez MA, Giulietti AM (1990) Effect of carbon and nitrogen source on growth and solasodine production in batch suspension of *Solanum eleagnifolium* Cav. Plant Cell Tiss Org Cult 21:55–60

Nishinari N, Syono K (1980) Changes in endogenous cytokinin levels in partially synchronized cultured tobacco cells. Plant Physiol 65:437–441

Ooms G, Hooykaas PJ, Moolenaar G, Schilperoort RA (1981) Grown gall plant tumors of abnormal morphology, induced by *Agrobacterium tumefaciens* carrying mutated octopine Ti plasmids; analysis of T-DNA functions. Gene 14(1–2):33–50

Otten L, Schilperoot RA (1978) A rapid micro scale method for the detection of lysopine and nopaline dehydrogenase activities. Biochem Biophys Acta 527:497–500

Parsons J, Rodríguez Talou J, Giulietti AM, Alvarez MA (2001) Solasodine production by undifferentiated and differentiated *Solanum eleagnifolium* Cav. *in vitro* cultures. Phyton Int J Exp Bot 70:159–163

Patel K, Singh RV, Patel DK (2013) Medicinal significance, pharmacological activities, and analytical aspects of solasodine: a concise report of current scientific literature. J Acute Dis 2:92–98

Regerat F, Pourrat A (1981) Isolement par voie fermentairs de la solasonine a partir d'un melange de glucoalcaloïdes. Planta Med 43:280–284

Rodriguez JA (1984) Producción de solasodina por hidrólisis microbiana de glicoalcaloides de *Solanum eleagnifolium* Cav. PhD thesis, Facultad de Química, Bioquímica y Farmacia, Universidad Nacional de San Luis

Rodríguez J, Segovia R, Guerreiro E, Ferretti F, Zamarbide De Sosa G, Ertola R (1979) Production of solasodine by microbial hydrolysis of glycoalkaloids of *Solanum eleagnifolium* Cav. and its application to the synthesis of 16-dehydropregnenolone acetate. J Chem Technol Biotechnol 29(8):525–530

Saini V, Midhha A, Gupta S, Rathore MS, Wate SP, Bhusari KP (2007) A simple spectrophotometric method for determination of solasodine in *Solanum xanthocarpum* capsule formulations. Internat J Plant Sci 2(1):128–129

Schreiber K (1963) Identifizierung vonβ 1-Tomatin als Hauptalkaloid in einigen experimentell erzeugten Mutanten von Lycopersicon esculentum Mill. *Solanum*-Alkaloide XXIX Mitteilung. Die Kulturpflanze 11(1):502–506

Schreuber K (1968) Steroid alkaloids: the *Solanum* group. In: Manske RMF (ed) The alkaloids-chemistry and physiology. Academic Press Inc., New York/London, pp 1–178

Shakirov R, Mirzaev YR, Bobakulov KM, Abdullaev ND (2012) Glycosylation of solasodine and pharmacological activity of the product. Chem Nat Compd 48(4):610–612

Spencer A, Hamill JD, Rhodes MJC (1990) Production of terpenes by differentiated shoot cultures of *Mentha citrata* transformed with *Agrobacterium tumefaciens* T37. Plant Cell Rep 8: 601–604

Suzuki T, Yoshioka T, Hara Y, Tamata M, Fujita Y (1987) A new bioassay system for screening high berberine-producing cell colonies of *Thalictrum minus*. Plant Cell Rep 6(3):194–196

Tambe NM (2013) Studies on in vitro callus culture of *Solanum khasianum* Clarke. Int J Recent Trends Sci Technol 5(3):111–112

Telek L, Delphin H, Cabanillas E (1977) *Solanum mammosum* as a source of solasodine in the lowland tropics. Econ Bot 31:120–128

Trivedi P, Pundarikakshudu K (2007) Novel TLC densitometric method for quantification of solasodine in various *Solanum* species, market samples and formulations. Chromatographia 65(3–4):239–243

Weiler EW, Kruger H, Zenk MH (1976) Radioimmunoassay for the determination of the steroidal alkaloid solasodine and related compounds in living plants and herbarium specimens. Planta Med 39:112–124

Weiler EW, Krüger H, Zenk MH (1980) Radioimmunoassay for the determination of the steroidal alkaloid solasodine and related compounds in living plants and herbarium specimens. Planta Med 39(6):112–124

Weston RJ (1976) An investigation of methods for isolation of the steroidal alkaloid solasodine from native New Zealand *Solanum* species. J Appl Chem Biotechnol 26:657–666

Zenk MH, El-Shagi H, Schulte U (1975) Anthraquinone production by cell suspension cultures of *Morinda citrifolia*. Planta Med S 01 (28) 79–110

Chapter 6
Molecular Farming in Plants

Abstract Molecular farming can generally be defined as the production of molecules (proteins, fatty acids) for the pharmaceutical and chemical industries in transgenic organisms (plants, animals, etc.).

Molecular farming in plants has advantageous aspects as biosecurity, they do not bear pathogens for humans or animals, and they do not produce toxins. Also, plant protein synthetic machinery is able to produce complex glycosylated proteins as they have a glycosylation pattern with slight difference respect of that of mammals, and can perform foldings.

When displayed in *in vitro* conditions, molecular farming have the characteristic advantages of that type of culture, mainly the capacity of working under Good Manufactory practices as is required by the pharmaceutical industry. The critical aspects of plants for molecular farming are the different glycosylation patterns respect to mammals, the duration of the productive process, and the relatively low yields.

The approval by the FDA of the first medicine to be used in humans, taliglucerase alfa (ELELYSO®), has give an impulse to this technology but there are some drawbacks to be addressed, particularly related to low yields and regulatory aspects.

Keywords Molecular farming • Recombinant proteins • Plant transformation • Plant glycosylation • *Agrobacterium tumefaciens*

6.1 Introduction

Plants have the unique ability of using solar energy and inorganic compounds to produce biomolecules, some of which are of medicinal interest, as was discussed in previous chapters of this book. Besides, using the available tools of genetic engineering, plants can produce foreign proteins. Molecular farming can be defined as the use of genetically modified organisms (plants, animals) to produce molecules of therapeutic or industrial interest. Although other molecules, as lipids, can be expressed, the term molecular farming is mainly applied to recombinant proteins.

M.A. Alvarez, *Plant Biotechnology for Health: From Secondary Metabolites to Molecular Farming*, DOI 10.1007/978-3-319-05771-2_6

6.2 Recombinant Proteins in Plants

The pharmaceutical industry requires recombinant proteins in sufficient amounts to cover the market demand and to bring a quick response to an epidemic outburst. There is an increasing requirement of those therapeutic proteins, which would not be satisfied in the next few years by the current productive platforms (bacteria, yeasts, or mammal cells). New platforms have demonstrated its potentiality to produce recombinant proteins, as transgenic animal and plants (Thomas et al. 2002; Joshi and López 2005).

The choice of a productive platform depends on the characteristics of the recombinant protein, the market needs, and the production costs among others factors. Each platform has its strengths and weaknesses. Mammal cells in general produce identical proteins as required, but have the risk of harbouring harmful agents (e.g.: prions, pathogens, oncogenes) requiring purification steps that increase costs. Bacteria, which have high production rates, cannot perform the complex post-translational processes typical of eukaryotes (glycosylation, folding) limiting the spectra of recombinant proteins they can express. Also, they require from extra purification steps since they store proteins in inclusion bodies or can produce endotoxins.

Plant cells, being eukaryotes, process proteins as mammals do but with slight variations (e.g., in the glycosylation pattern), and are able to produce complex proteins, including multimeric and glycoproteins (Joshi and López 2005; Stoger et al. 2005; Mascia and Flavell 2004; Daniell et al. 2009). Besides, plants do not harbour pathogens, or produce endotoxins, or include proteins in vesicles. The production of recombinant proteins in plants can be performed in agronomic cultures, in the field, or in *in vitro* cultures. In the field, scaling up can be achieved by increasing the cultured area with the potential production of hundreds of kilograms of purified protein per year. In the case of *in vitro* cultures, the scaling up can be achieved increasing the volume of the culture. The disadvantages of plants for producing recombinant proteins are the long extent of the process, the variation of yields and quality of the product, and the difficulty to use good manufacturing (GMP) and good laboratory practices (GLP) in the early stages of production in the field. In addition, there are susceptible to pests and diseases, or can be contaminated by agrochemicals or fertilizers. Furthermore, there are risks of gene transfer to conventional cultures, which requires working under specific regulations for GMOs in the field (Obembe et al. 2011; Xu et al. 2012).

Producing recombinant proteins in *in vitro* cultures has the characteristic advantages of the system (Chap. 4) as well as the capability of producing proteins in GMP and GLP (Hellwig et al. 2004). Also, scaling-up and product purification are simple (Medina-Bolivar and Cramer 2004; Medina-Bolivar et al. 2003).

More than 50 % of the total production cost of a process is related to the extraction and purification of the proteins, which makes critical the operations of down stream processing (DSP). In *in vitro* cultures, the DSP costs can be remarkably reduced if the protein of interest is secreted into the culture medium (Hellwig et al. 2004; Streatfield and Howard 2003; James et al. 2002; Faye et al. 2005).

A bottleneck for the commercial exploitation of plants for producing proteins is their low productivity. To be competitive, the plant system must be optimized at the gene expression, culture development or downstream processing level (Xu et al. 2012).

6.3 Plant Transformation

Plant transformation can be achieved by using direct or indirect natural procedures (Fig. 6.1). Usually, the gene that code for the protein of interest is introduced into the plant genome in a construct that also contain a promoter, a terminator, and different signals that can modulate the expression of the gene in intensity, time and space.

6.3.1 *Direct Methods*

The most common methods are the physical methods that imply the use of biolistic and electroporation. Chemical methods are PEG, calcium phosphate, and artificial lipids-mediated, among others (Fig. 6.1). Biolistics is one of the most popular direct methods, e.g.: for chloroplast transformation (Bock 2014), for constitutive or in specific-organ expression of the recombinant proteins (Cunha et al. 2011a). These simple methods have the disadvantage of a high rate of cellular damage, and the increased rate of silencing due to multiple copy integration (Egelkrout et al. 2012).

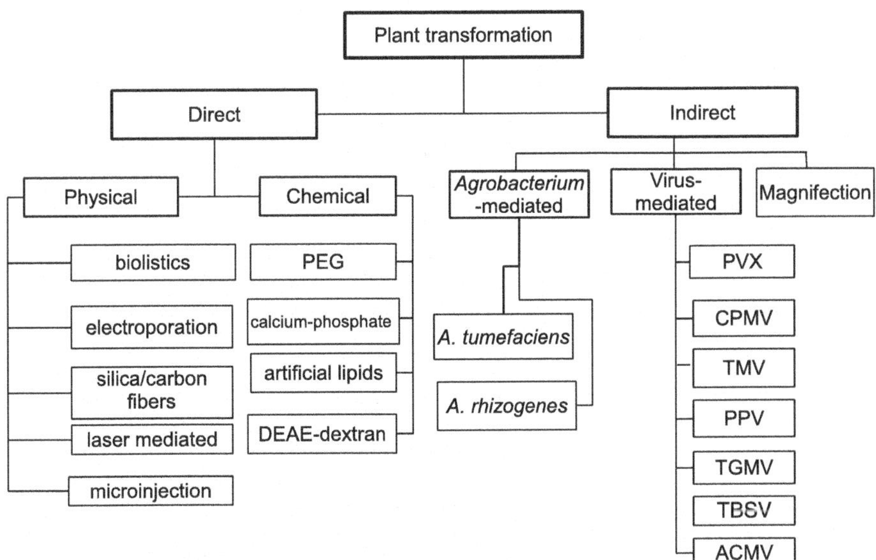

Fig. 6.1 Plant transformation methods

6.3.2 *Indirect or Biological Methods*

Indirect methods involve the transference of the DNA using *Agrobacterium* or viral vectors.

6.3.2.1 *Agrobacterium*

Agrobacterium is a natural bacterium with the capability of infecting Dicotiledoneae plants producing the tumoral disease known as the "crown gall" disease (de la Riva et al. 1998). During the infection a fragment of a plasmidic DNA (T-DNA) is transferred to the nucleus of the plant cell. That T-DNA contains two types of genes, the oncogenes (codifying for enzymes involved in the synthesis of auxins and cytokinins responsible for the tumour formation), and the genes codifying for the synthesis of opines. Opines (a condensation of amino acids and sugars) are synthesized and secreted by the gall crown cells, and used by *A. tumefaciens* as carbon and nitrogen source. Out from the T-DNA are located the genes for the catabolism of opines, the genes involved in the transference of the T-DNA from the bacteria to the plant cells (*vir*), and the genes involved in the conjugation among bacteria. The T-DNA enters into the plant and integrates randomly in the plant genome by illegitimate recombination. Once integrated, the expression of the oncogenes and of the genes that induce the production of opines is triggered (de la Riva et al. 1998). As a consequence, there is an uncontrolled growth of plant cells an outgrowth of bacteria.

6.3.2.2 The Plasmid Ti

The tumour inducing plasmids (plasmid Ti) are classified according to the induced opines (de la Riva et al. 1998). The most studied are the nopaline/agrocinopine (e.g.: pTiC58 y pTiT37) and the octopine/manopine (pTiB6 y pTi15955). The T-DNA is flanked by 25 bp inverse repeated sequences, the right and left borders (RB and LB respectively) that act as *cis* elements for the transference apparatus. There is a polarity established between them being the RB the first to enter into the plant cell (Gelvin 2003). A remarkable finding was that any foreign DNA placed between the RB and LB can be transferred to the plant cell no matter its origin. That has allowed the construction of vector systems for plant transformation (de la Riva et al. 1998).

6.3.2.3 Vectors

(a) Binary vectors: The introduction of foreign genes in the T region of the *A. tumefaciens* plasmid Ti has some difficulties. First, the plasmid Ti lacks of unique recognition sites for restriction enzymes. Secondly, there is a limit in the size

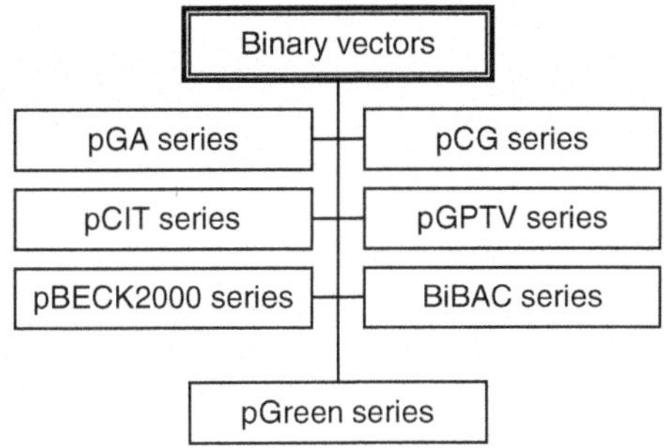

Fig. 6.2 Main binary vectors in use for plant transformation

the plasmid can bear; DNA fragments of high molecular weight complicate the manipulation (Gelvin 2003). Numerous vectors derived from the disarmed Ti plasmid have been developed (An 1985). In binary vectors two different plasmids are employed, a small plasmid, that has origins of replication both for *Agrobacterium* as *for Escherichia coli*, and do not have its tumour-inducing genes. The plasmid also contain unique restriction sites in the T-region allowing the introduction of the desired genes, promoter sequences, selection markers (antibiotic resistance), directional signals, etc., and are flanked by the right (RB) and left (LB) borders of the Ti-DNA (Gelvin 2003; Hellens et al. 2000). The second plasmid is a helper plasmid that harbours the *vir* genes but do not have the T-DNA region. The main binary vectors are from the series pGA, pCG, pCIT, pGPTV, pBEK2000, binary-BAC (BiBAC) and pGreen (Fig. 6.2).

Each series of vectors has a particular characteristic that makes them useful for plant transformation, e.g., antibiotic resistance genes, particular cloning sites, etc.

Briefly, vectors from the (a) pGA series contain a RK2 ori genes that confer resistance to tetracycline and kanamycine and G418 (*nptII*) and a polylinker site, (b) pCG series contain the pRiHRi (*A. rhizogenes* root-inducing) ori gene, and a *ColE1* ori, (c) pCIT series contain the hygromycin (*hph*) resistance gene and the *cos* site for cloning long DNA sequences, (d) pGPTV series contain marker genes close to the LB, (e) pBECK2000 with synthetic T-DNA borders, genes for phosphinothricin resistance (*bar*), the *p1 Cre/loxP* site recombinase system that allows site-specific excision of marker genes after transformation and the transfer and integration of foreign genes as a single or different T-DNA units, (f) Binary-BAC (BiBac) series, based in the bacterial artificial chromosome vector, low-copy numbers of origenes, a helper plasmid with additional virgenes copies, the elements of the vector are specific to allow transformation with DNA of high molecular weight, (g) pGreen series, contains the ori *pSa*, the *ColE1* ori (from pUC), a *rep A* gene, and multiple cloning sites based on pBlue-Script vector.

(b) Gateway® vectors: the Gateway® cloning system permitted the construction of more rapid and efficient binary vectors. This technology was initially based on phage lambda site-specific recombination (Hartley et al. 2000; Karimi et al. 2002). The phage genome integrates into *E. coli* genome by recombining its *attP* site with the *attB* site of the bacteria (BP reaction); as a result phage genome is flanked by the *attL* and *attR* sites. The reaction is reversible (LR re-action). The manufacturer provides the enzymes needed for performing both reactions: BP clonase (phage integrase, Int, plus *E. coli* integration host factor, IHF) and LR clonase (Int, IHF plus excisionase, Xis). The basis of Gateway® cloning method is the production of a recombinant DNA by means of the *att* sites and the clonases. There are numerous binary vectors compatible with the Gateway® cloning methodology that can be used as destination vectors (Karimi et al. 2007). The Multisite vectors further developed permitted the recombination of DNA fragments within certain size limits in a single recombination step (Cheo et al. 2004; Karimi et al. 2005; Magnani et al. 2006). Some Gateway compatible binary vector series (pGWB) were made with intermediate plasmids pUGWs, based on pBI plasmids, or pPZP plasmids. For promoter swapping a R4 Gateway binary vector (R4pGWB) series was constructed (Tanaka et al. 2012).

The MultiRound Gateway technology, with two different entry vectors that can sequentially deliver multiple DNA fragments into a Gateway-compatible destination vector, allow stacking multiple DNA fragments (Chen et al. 2006). An improved version of the last one was developed introducing a transformation-competent artificial chromosome-based destination vector and a recombination deficient *Agrobacterium* strain. The resultant vector let introduce up to eight genes placed on a single destination vector, besides other sequences as selection markers, scaffold attachment regions, etc. The inserted transgenes were stably expressed for at least two generations (Buntru et al. 2013).

(c) Co-integrated vectors: they are not commonly in use because they require of long homologies between *Agrobacterium* and *E. coli* being less efficient than binary vectors. They are engineered using a Ti-plasmid (SEV-series with a kanamycine resistance gene, pGV series, with a segment of pBR322), intermediate vectors to allow the transference of high-molecular weight DNA sequences, and a helper vector. The resulting vector contains vir genes, a foreign DNA sequence, selectable markers to plants and bacteria, and LB and RB.

6.3.2.4 Viral Vectors

Viral vectors are attractive for their ease and simple manipulation. They are classified in (a) epitope presentation systems, with the peptide displayed in the viral surface, and (b) polypeptide expression systems, with the transgene usually expressed as a non-fused polypeptide in infected cells. In the first group are vectors derived from the cowpea mosaic virus (CPMV), tobacco mosaic virus (TMV), tomato bushy stunt virus (TBSV), plum poxvirus (PPV), and alfalfa mosaic virus (AIMV). In the second group, are vectors derived from TMV, potato virus X (PVX), PPV, TBSV, CPMV (Cañizares et al. 2005). Also, viral vectors were developed based on the geminivirus tomato golden mosaic virus (TGMV) (Hong et al. 2004).

6.3.2.5 Promoters

Gene regulation is mediated by signals placed contiguously (*cis*-acting) or not (*trans*-acting) to the encoding-sequences. Promoters generally appertain to the first category of modulators; usually they are binding sites for transcription factors (TFs), which are trans-acting modulators of gene-expression. They can be only active in a particular species, genus or family (homologous) or in different species or even kingdoms (heterologous).

For plant transformation, promoters generally in use can be (a) constitutive, generally heterologous, which direct expression in the whole plant without being conditioned by exogenous or endogenous factors, (b) tissue specific, usually homologous, that drive gene expression to a particular tissue or stage of plant development (e.g., roots, seeds, parenchyma, etc., (c) inducible, which are conditioned by exogenous stimuli, as biotic and abiotic factors (light, temperature, oxygen level, chemicals, wounding) allowing gene modulation by antibiotics, metals, herbicides, etc., and (d) synthetic promoters, tailored with elements of different promoter regions (Fig. 6.3).

Other regulatory systems are based on trans-activating proteins that can be used in combination with promoters. In eukaryotes, promoters contain several elements, some of them present in the majority of promoters: the CAAT or the AGGA box, a consensus sequence −80 bp from the start point, that increase the promoter strength; the TATA box, placed 25 bp upstream of the start point, usually surrounded by GC rich sequences, that is the place of binding of RNA polymerase II and some TFs; a GC box, rich in guanidine (G) and cytosine (C), usually found in multiple copies in the promoter region, in general surrounding the TATA box; and a CAP site, which is the transcription initiation sequence (+1). Out from the promoter sequence, enhancer sequences can be added upstream or downstream-from the promoter, they enhance promoter activity by binding some TFs, also, in some cases; they can confer tissue specific or stage-specific gene expression (Fig. 6.3).

(a) Constitutive promoters: they are the most popular promoters at least for research. The main reasons are that they are strong promoters, which mean that

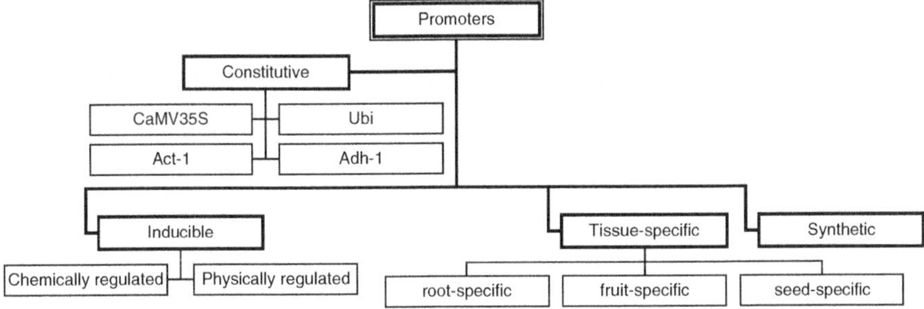

Fig. 6.3 Classification of promoters for plant transformation

the level of protein expressed is usually high, and that the distribution of the protein expressed is ubiquitous during all stages of plant development.

- CaMV 35S: promoter of the Cauliflower Mosaic Virus is a very strong promoter, highly efficient in Dicotiledoneous.
- Ubi: promoter that controls the expression of the *Ubi-1* and *Ubi-2* genes encoding ubiquitin in maize. A relevant region contained in *Ubi* is two overlapping sequences, heat shock inducible, located –214 to –204 from the start site.
- Act-1: in the 5′ region of rice actin-1 gene it has two relevant regions, a 38 bp poly (dA-dT) element located from −245 to −152, which is the place of binding for trans-acting positive regulators of Act-1, and a CCCAA pentamer repeats, from –300 to –260, involved in the negative regulation.
- Adh-1: promoter of maize alcohol dehydrogenase. The anaerobic regulatory elements (ARE) of the promoter and the first intron of Adh-1 gene sequence are combined in one promoter named EMU that also contains enhancer elements from the octopine synthase (OCS) gene from *A. tumefaciens*, and a TATA box.

(b) Tissue-specific promoters: they control gene expression in a tissue-dependent mode or according to the plant development phase. There are numerous inducible promoters in use. In general they are homologous since the modulation must be carefully adjusted. Some examples are for:

- Seed-specific: beta amylase promoter, barley hordein gene promoter.
- Root-specific: tobacco RD2 gene promoter.
- Ovary-specific: tomato pz7 and pz130 gene promoters.
- Fruit-specific: banana TRX promoter, melon actin promoter.

(c) Inducible promoters: their main characteristic is that they allow that the expression of a particular gene (or genes) can be turned on or off under certain conditions. They can be chemically or physically regulated.

Chemically regulated: their activity is regulated by different chemicals as alcohol, tetracycline, metals, and steroids. Those chemicals are not plant metabolites, but must be specific for the gene under study, no toxic for the plant and easily manipulated. Into this group are the pathogenesis regulated (PR) promoters, induced by salycilic acid (SA), ethylene and benzothiadiazole (BTH), chemicals that are liberated by the plant during pathogenesis. Usually BTH, which act systemically and has a long-term effect, is best tolerated for plants than (SA), which acts in the site of application.

Physically regulated: they are induced by environmental factors as temperature (cold or heat-shock induced), light (e.g.: from pea or myxobacterium), salt stress, water availability, etc. Some examples are the light-responsive elements in the ribulose-1,5-biphosphate carboxylase-oxygenase and chalcone synthase gene promoters.

(d) Synthetic promoters:

They are tailor-made by combining sequences from other promoters they must contain at least a TATA box, a CAP site, and a CCAAT consensus sequence.

6.3.3 *Stable Transformation*

In 1983 Herrera-Estrella et al. reported the functional expression of a foreign gene after engineering a T-DNA replacing the *vir* genes for the gene of the heterologous protein. The construct also carried a promoter, a terminator, and a gene that codified for a selection marker. That was the starting point for expressing recombinant proteins using *Agrobacterium* as a vector. Since then, a variety of technologies have been developed to attain the expression of genes in plants with high yields (Fig. 6.4).

6.3.3.1 Nuclear Transformation

The gene for the protein of interest is usually under the control of a strong constitutive promoter, as the promoter of the cauliflower mosaic virus (CaMV35S). In that case the protein is expressed in the whole plant but, usually is extracted from leaves. In the cell, the recombinant protein can be targeted to a specific sub-cellular location, as the endoplasmic reticulum, the cytoplasm, etc., by the addition of the proper signals.

Tobacco is one of the more employed species, its transformation is not complicated, produces a high amount of biomass (more than 30 t per acre), and is not in the food chain (Tremblay et al. 2010). Besides, several tobacco varieties with low nicotine and alkaloid content have been developed, which overcome the potential toxicity of the products (Menassa et al. 2001). Other species with good leaf development as *Lactuca sativa*, *Medicago sativa*, *Trifolium* sp. are also used. Two limitation of expressing recombinant proteins in leaves are their short shelf life, which requires an immediate post-harvest processing to assure the stability and quality of the protein, and the susceptibility to environmental factors (biotic and abiotic), which may affect the yields.

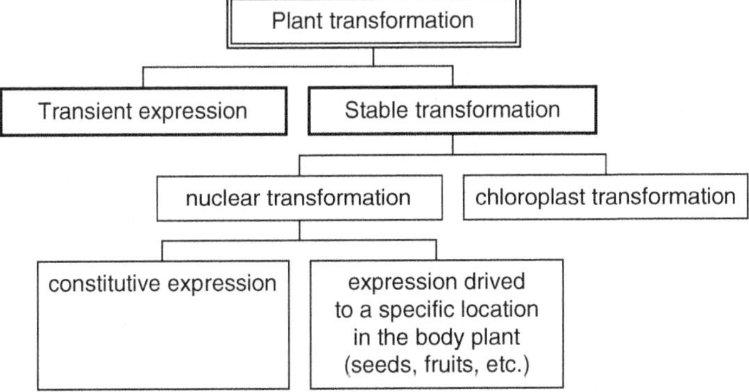

Fig. 6.4 Proteins can be expressed in plants in two ways: transiently (e.g. by Agroinfiltration), or by stable transformation using the protein synthesis machinery of the nucleus or chloroplasts

Fruits, as tomato (*Solanum lycopersicum*) has also been used to produce PMPs (Mason et al. 2002), e.g.: an antigen fusion protein, F1-V, against *Yersinia pestis*. Hyroponic transgenic cultures were grown in a greenhouse with continuous harvest of fruits. The estimated maximal antigen production was about 293 mg m^{-2} per year, which can be improved ameliorating the environmental conditions and culture practices (Matsuda et al. 2009).

The recombinant proteins constitutively expressed in plants include antibodies, vaccines, citoquines, growth hormones and industrial proteins (Benchabane et al. 2008; Sharma and Sharma 2009) (Table 5.1). The first clinical assay for a recombinant protein was a variant of the Guy's13 secretor antibody expressed in tobacco (Ma et al. 1998).

6.3.3.2 Chloroplast Transformation

Gene integration, by homologous recombination, in chloroplasts produces hundreds of copies of the transgene and consequently high recombinant protein yields were attained (Bock 2014; Gao et al. 2012; Day and Goldschmidt-Clermont 2011; Daniell et al. 2005; Gleba et al. 2005). Chloroplasts perform relevant post-translational modifications, the system is devoid of gene silencing, and has the advantage of the maternal inheritance that reduces the possibility of transgene transmission through pollen, and allows to stack multiple transgenes in operons and co-express them from a single promoter as a polycistronic mRNA (Bock 2014; Thyssen et al. 2012).

Some drawbacks are the expression limited to non-glycosylated proteins, protein instability, and the low expression levels in non-green tissues.

Enzymes, antibodies, antigens and other proteins of interest for the pharmaceutical industry (e.g.: human serum albumin) have been expressed in plastids (Fernández-San Millán et al. 2003; Clarke et al. 2011; Bock 2014).

6.3.3.3 Expression in Seeds

The main advantage of expressing proteins in seeds is their stability during storage at room temperature with a minimal lack of protein activity (Boothe et al. 2010). In this case, protein expression is directed to seeds through the appropriate sequence-signal. In seeds, proteins, protected from proteolytic degradation, attain high expression levels (7–10 % TSP). Most of the seeds used to express recombinant proteins belong to species in the food chain (Oryza sativa, *Triticum aestivum*, *Glycina max*, *Zea maïs* and *Hordeum vulgare*), which demands strict protocols and documenting each step of the process (Ramessar et al. 2008; Ramessar et al. 2009). Some of the proteins expressed in this way are the functional recombinant human growth hormone and the coagulation factor IX in soybean seeds (Cunha et al. 2011a, b).

The extraction of proteins from seeds has been improved by the fusion of the recombinant protein to oleosines that accumulate in oil bodies. This technology has allowed the production of recombinant human insulin in transgenic safflower (Univ. of Calgary).

6.3.4 Transient Expression

Proteins can be transiently expressed, with no need for stable integration of the transgene. The foreign genes are delivered by vacuum infiltration in the mesophylle of leaves. The vectors can be *Agrobacterium* spp. (Kapila et al. 1997), viral vectors (Verch et al. 1998), or a combination of both (magnification) (Gleba et al. 2005). In general, the production of recombinant proteins started 24 hs post-infection and continues for several days (*Agrobacterium* mediated) to several weeks (virus mediated) (Gleba et al. 2005).

Initially, transient transformation was used to test the efficiency of the constructs designed for expressing genes and for validating the activity of newly expressed recombinant proteins (Circelli et al. 2010; Whaley et al. 2011). Then, it became attractive for producing biopharmaceuticals due to their high protein yields and to not generate a GMO (Circelli et al. 2010; Sheludko 2008). Another advantage of this system is that it allows the rapid co-expression of multiple genes (Huang and Mason 2004; Medrano et al. 2009).

Transient expression was performed in several species, e.g.: *N. benthamiana*, *Arabidopsis*, *Solanum tuberosum*, *Phaseolus*, *Lactuca sativa*, etc.

Some authors have found that with *Agrobacterium* yields was between 0.1 a 180 $\mu g\ g^{-1}$ (Medrano et al. 2009). In general, protein yields are between 4 to 20-fold higher than by stable transformation (using the CaMV35S promotor) (Medrano et al. 2009). With viral vectors the yields are higher than with *Agrobacterium*, e.g.: 5 $mg\ g^{-1}$ for GFP (green fluorescent protein) (Marillonnet et al. 2004). However, the expression usually is after the 14th day of infection, which complicates the stability of the recombinant protein.

Magnification is a shorter process, with quite higher protein yields, versatile, with no wild type virus generated (Gleba et al. 2005). By agroinfiltration, the HIV-1 *Nef* protein was expressed in *N. benthamiana* leaves testing also the effect of gene-silencing viral suppressors on yield (Circelli et al. 2010). The highest *Nef* yield (1.3 % of total soluble protein) was attained when the P19 of Artichoke Mottled Crinckle virus was co-expressed, which correlates with a fall in the amount of Nef-interfering RNAs.

An interesting approach was the use of a transient large-scale plat-form to a fast reaction to a pandemic outbreak of influenza (D'Aoust et al. 2010). The H5 virus-like protein (VLP) was expressed by agroinfiltration in *A. thaliana* at large-scale (1,200–1,500 plants per week), obtaining approximately 25 kg of leaf biomass at the 6th day post-infiltration. Using this technology the research group, from the Medicago Inc. company, has tested an experimental vaccine against a novel strain (A/H1N1) that appeared in 2009, proved its immunogenicity in mice, and demonstrated its efficacy and speed of response (Landry et al. 2010; D'Aoust et al. 2010) with clinical trials are in progress. In 2013 they have developed a plant-based Rotavirus VLP vaccine candidate using the same platform. Some companies are using this platform for producing commercial proteins, e.g. Genaware produces α-tricosantine and acid lyzosomal human lipase, MagnIcon produces hepatitis B antigen and viral Norwalk-like particles virus and F1 and V *Yersinia pestis* antigens, and by Geminivirus technology the heavy and light chains of the monoclonal antibody 6D8 IgG–like against the Ebola GP1 virus are produced. The Launch expression system combines the genome of a plant viral vector with the *Agrobacterium* binary plasmid. The expression of the recombinant influenza H5HA-1 protein attained a yield of 60 μg g^{-1} of fresh leaves (Yusibov and Mamedov 2010).

Those systems are used at industrial scale in the Fraunhofer Center for Molecular Biotechnology (Newark, DE), Medicago Inc (Quebec, Canadá), Texas AandM (College Station, TX) and Kentucky BiProcessing LLC (Owensboro, KY). Terrasphere have developed a high-density hydroponic system that can produce high amounts of recombinant proteins.

6.4 *In vitro* Cultures

In vitro cultures (suspended cells, hairy roots) have the advantage of managing the process in controlled environmental conditions under GMP and GLP. Production is faster than in agronomic cultures, and down stream processing is easier and cheaper specially when the product is recovered from the culture medium. Production can be performed in undifferentiated (suspended cells) or differentiated (hairy roots) cultures. A variety of recombinant proteins have been expressed such as antibodies, antigens, enzymes, etc.

6.4.1 Cell Suspension Cultures

Numerous proteins have been expressed, mainly in *N. tabacum* strains BY-2 (Bright Yellow-2) and NT-1 (*N. tabacum*-1) which are easily transformed with *Agrobacterium*, have a fast and robust growth, with the ability of synchronizing their cell cycle. Also, *N. tabacum* cv. Xanthii, carrot, rice, etc. have been used (Terashima et al. 1999). The yields or recombinant protein attained are variable, the low levels were attributed to protein degradation by proteases or the adsorption of proteins onto the surfaces of the culture vessels (Doran 2000). Other proteins produced in cell suspension cultures arte factor XIII-A domain in tobacco (Gao et al. 2004), human interleukins in tobacco (Magnuson et al. 1998; Kaldis et al. 2013). Besides tobacco, *N. benthamiana*, *Arabidposis thaliana*, rice, carrot, and other species have been used for producing recombinant proteins in cell suspension cultures.

6.4.2 Hairy Roots

Hairy roots are obtained by infecting plants with *A. rhizogenes* (see Chap. 4). *A. rhizogenes* bears the Ri plasmid that induces root formation in the plants that infects. The integration of the T-DNA plasmid in the plant genome results in a differentiation and development of neoplasic roots, hairy roots, in the infection sites. Hairy roots are characterized by an indefinite growth with high genetic stability. Hairy roots, cultured in controlled conditions, produce recombinant proteins in GMP. Besides, it is possible the extracellular secretion or rhizosecretion of the protein, allowing a less expensive recovery in a well defined culture media with low protein content.

Recombinant antibodies, enzymes, antigens, growth factors, immunomodulators, interleukine-12, human acetylcholinesterase, etc. have been expressed in hairy roots (Wongsamuth and Doran 1997; Woods et al. 2008).

A limitation of hairy roots is the filamentous, highly branched morphology that difficult their culture in classical bioreactors. To overcome that inconvenient, new bioreactor designs were developed. An example is the mist reactor that offers a low hydrodynamic stress with high volumetric oxygen transference levels (Weathers and Giles 1988).

6.5 Plant Glycosylation

Plants can manage protein synthesis performing modifications in a similar way than mammal cells with several differences although the sequence-specificity is not always conserved (Gomord and Faye 2004). Also, N-glycosylation differences could represent a severe limitation on the use of plant-made pharmaceuticals that triggered abundant work on the field (Bardor et al. 2006), O-glycosylation, which has been less studied (Daskalova et al. 2010).

N-glycosylation usually affects protein stability, solubility, folding, and biological activity (Bosch et al. 2013). The first steps of protein synthesis, that take place in the endoplasmic reticulum, are similar in plants as mammals; the oligosaccharide precursor on the nascent protein backbone is transferred to a specific residue of asparagine. The protein follows its path through the ER and Golgi where suffers the removal or addition of sugar residues. The process in plants and mammals differs when the nascent proteins is transferred to the Golgi apparatus. In plants, a β-1,2-xylose is linked to a core mannose residue, and a α-(1,3) fucose is linked to the proximal N-acetylglucosamine. Also, plants lack the terminal β-(1,4) galactose residues (Fig. 6.5). In general, plant-derived proteins are a heterogeneous N-glycan group, with a few exceptions, e.g., the monoclonal C5-1 antibody produced in alfalfa (Bardor et al. 2003; D'Aoust et al. 2004).

Plants have been engineered in several ways to generate recombinant glycoproteins with a customized N-glycosylation pattern (Castilho et al. 2011). The enzymes responsible of the addition of residues typical of plants were knocked out, or the expression of the human glycosyltransferase and the introduction of sialic acid were engineered (Castilho et al. 2012) for their effect on protein biological activity and half-life (Abranches et al. 2005).

Additionally, N-glycosylation can be modified *in vitro*, e.g.: the β-(1,4) -glycosyltransferase from bovine milk was added to transgenic alfalfa plants (Bardor et al. 2003) and also sialic acid (Raju et al. 2001).

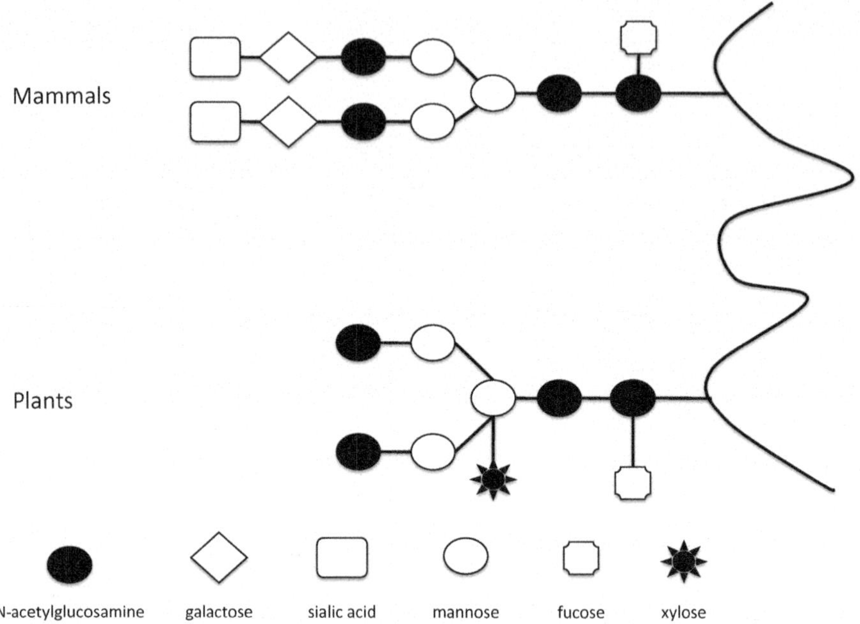

Fig. 6.5 Glycosylation pattern of plants and mammals

In addition, gene disruption by homologous recombination was performed to design a double knockout strain that lacks the two plant-specific sugar residues, fucose and xylose, implicated in allergic immune responses (Lerouge et al. 2000; Koprivova et al. 2004).

As for O-glycans, they are primarily attached to hydroxyproline and serine residues in plant glycoproteins. In mammalian cells, O-glycans are mucin type O-linked glycans, which are synthesized in the Golgi in a succession of steps, independently of the consensus motifs. It was demonstrated that plants can be engineered to express a complex O-glycoprotein, LTBMUC1 (*E. coli* enterotoxin B subunit: *H. sapiens* mucin1 tandem repeat-derived peptide fusion protein), with the mammal characteristic O-glycosylation. LTBMUC1 was expressed in *N. benthamiana* by stably and transient transformation using *A. tumefaciens* that also co-express human GalNAc-T2, the *Caenorhabditis elegans* UDPGlcNAc/UDPGalNac transporter, and *Yersinia enterocolitica* UDP-GlcNAc 4-epimerase required for the GAlNAc-glycosylation typical of mammal cells. LTBMUC1 also appeared decorated with plant-specific glycans attached to the hydroxyproline residues (Daskalova et al. 2010). Recent studies have shown that plant-specific Hyp-O-glycosylation might be an alternative to PEGylation to increase the serum half-life of human therapeutic proteins; however, it is necessary to perform an analysis of the possible induction of immunogenicity and/or allergenicity in humans (Gomord et al. 2010).

6.6 Plant-Made Proteins for Health

There are numerous reports about expression of recombinant proteins in plants (Ahmad et al. 2012). The most used species are *N. tabacum*, *A. thaliana*, *Oryza sativa*, tomato, *Medicago sativa*, *Glycina max*, *S. tuberosum*, *Zea maïs* (Twymann et al. 2003; Fischer et al. 2004). Technical advances have permitted the transformation of a large variety of plant species including non-conventional species as *Medicago truncu-lata*, which produces more homogeneous glycans structures (Abranches et al. 2005).

6.6.1 Antibodies

Antibodies are one of the more relevant groups of biopharmaceuticals, with a growing demand that might rapidly not be covered by the platforms in use. Hiatt et al. (1989) reported the expression of an antibody in plants for the first time. Initially, the heavy (H) and light (L) chains of the immunoglobulin were cloned separately. Then the resultant transformed plants were sexually crossed; the progeny contained some transgenic plants that expressed the assembled chains as a functional antibody. Later on, the cloning of the genes for both chains in a single event was reported, producing a plant that expresses the whole functional antibody (Table 6.1). Functional antibody fragments are also expressed, the fragment antigen binding (Fab), containing one of the two identical combining sites of the immunoglobulin, the antibody single bivalent fragments F(ab')2 containing the two combining sites, fragments consisting of only the variable regions of the L and H chains (Fv), that retain the antigen binding capacity, single-chain variable fragments (scFv) consisting of the VL and VH domains linked by a flexible peptide, maintaining the binding capacity (Seddas-Dozolme et al. 1999; Eeckhout et al. 2004; Galeffi et al. 2005), and single-domain antibodies (VHH) from camelids (Marconi and Alvarez 2014). Secretory antibodies (IgA) are also expressed (Larrick et al. 2001).

Table 6.1 Selected examples of antibodies expressed in plants

Antibody	Plant species	Reference
C5-1	*Medicago sativa*	Bardor et al. (2003)
sIgA anti-*S. mutans*	*N. tabacum*	Ma and Hein (1995)
Human anti-Rhesus D IgG_1	*Arabidopsis*	Bouquin et al. (2002)
Anti-CD4 and anti-CD28 scFv	*Triticum estivum (cv. Westonia)*	Brereton et al. (2007)
MAK33	*N. tabacum, S. tuberosum*	De Neve et al. (1993), De Wilde et al. (2002)
Guy's 13	*N. tabacum*	Wongsamuth and Doran (1997)
C5-1	*N. benthamiana*	Sainsbury and Lomonossoff (2008)
LO-BM2	*N. tabacum*	De Muynck et al. (2009)
Caro-Rx	*N. tabacum*	Fischer et al. (2003)

6.6.2 Vaccines

Plants also express immunogen antigens that will be used for formulating experimental vaccines (Table 6.2). The production of plant edible vaccines was proposed in the late '1980s, but it took several years to finally prove the concept (Haq et al. 1995; Mason and Arntzen 1995). Currently, there are some vaccines with plant made antigens in its formulation in clinical trials (Ahmad et al. 2012; Franconi et al. 2010), to be used as an edible, parenterally or topically administered vaccine (Giorgi et al. 2010).

The generation of transgenic plants by plant nuclear transformation has successfully produced mucosal vaccines against cholera, Norwalk virus, hepatitis B, and foot-and-mouth disease. The cholera toxin B subunit, the first plant-derived vaccine arising from plastid transformation, accumulates to 4.1 % total soluble protein in tobacco leaves (Daniell et al. 2001; Walmsley and Arntzen 2003).

Also, the diphteria toxin (DT), tetanus fragment-C (TetC) and the non-toxic S1 subunit of pertussis toxin (PTX S1) antigenic proteins have been expressed in low-alkaloid tobacco plants and carrot cell cultures. The coding sequences were placed, under the control of the strong promoters RbcS1 (tobacco) or CaMV35S (carrot), into the binary vector pBINPlus having the *nptII* gene for kanamycine selection. The antigens were produced in tobacco leaves or in optimized carrot cell suspension cultures, after extraction and purification they were used to formulate a vaccine that was injected to BALB/c mice. A strong antigen-specific serum antibody response was triggered after exposition to a commercial native diphteria, tetanus or pertussis toxoid (Brodzik et al. 2009).

Tomato was tested as an edible vaccine against malaria (Chowdhury and Bagasra 2007), lettuce against the *E. coli* heat-labile toxin B subunit (Martínez-González et al. 2011), rice against the roundworm *Ascaris suum* (Matsumoto et al. 2009).

Table 6.2 Selected examples of antigens expressed in plants

Recombinant antigen	Plant species	Reference
Hepatitis B surface antigen	*N. tabacum*	Kostrzak et al. (2009)
	S. tuberosum	
Hepatitis B surface antigen	Cherry, tomato	Gao et al. (2003), Thanavala et al. (2005)
TetC (Tetanus vaccine antigen)	*N. tabacum*	Tregoning et al. (2004)
Bacillus anthracis protective antigen	*N. tabacum*	Watson et al. (2004)
Hepatitis B surface antigen	Banana	Kong et al. (2001)
Hepatitis B surface antigen	*N. tabacum*	Kostrzak et al. (2009)
Anthrax protective antigen	*N. Tabacum*	Brodzik et al. (2009)
	Daucus carota	
DTP subunit vaccine	*M. sativa*	Brodzik et al. (2005)
ESAT6:Ag85B tuberculosis antigens	*A. thaliana*	Dorokhov et al. (2007)

A complete review about plant-made vaccines was published by Awale et al. (2012). Information about the use of plant-made immunogens for developing veterinary vaccines is presented in Chap. 7.

6.6.3 Other Pharmaceutical Proteins

Besides antigen and antibodies, protein allergens, enzyme, enzyme inhibitors, coagulation factors, cytokines, and hormones were expressed in plants (Benchabane et al. 2008) (Table 6.3). Peptides, as the synthetic antimicrobial peptide D4E1 were expressed in tobacco (Cary et al. 2000). Crude extracts resulted inhibitory for *Aspergillus flavus* and *Verticillium dahliae*, while transformed plants were resistant to *Colletotrichum destructivum*. Also, the HPV16 E7 and HPV16 L2 peptides were transiently expressed in *N. benthamiana*. A synergistic interaction was found using viral PVX-based expression vector pGR106 with an increase in the level of the recombinant peptides expressed (Cerovská et al. 2008).

Human somatotropin has been observed to accumulate to 7 % total soluble protein in plastid transformation, 300-fold greater than in nuclear-transformed tobacco (Staub et al. 2000), whereas human serum albumin accumulates to 11.1 % total soluble protein, 500-fold higher than in nuclear transformed leaves (Fernández-San Millán et al. 2003).

The human coagulation factor IX (hFIX) was expressed, using biolistics, in *Glycine max* seeds attaining amounts up to 0.23 % (0.8 g kg^{-1} seeds) of total soluble protein content. The recombinant (hFIX) remained stable and functional for up to 6 years stored at room temperature (Cunha et al. 2011a).

Table 6.3 Selected examples of recombinant proteins expressed in plants

Recombinant antigen	Plant species	Reference
Serum albumin	*N. tabacum*	Peter et al. (1990)
α-Amylase	*M. sativa*	Austin et al. (1995)
Human Protein C	*N. tabacum*	Cramer et al. (1996)
LT-B (heat-labile toxin B)	*Zea maïs* kernels	Chikwamba et al. (2003)
Avidin	*Zea maïs*	Hood (2002), Hood et al. (2002)
Human lactoferrin	*N. tabacum*	Jeffrey et al. (2000)
Tyrosin	*Zea maïs*	Woodard et al. (2003)
α1-antitrypsin	*Oryza sativa*	McDonald et al. (2005)
Tricosanthin-α	*Trichosanthes kirilowii*	Lei et al. (2006)
Hirudin from *Hirudo medicinalis*	*Canola*	Demain and Vaishnav (2009)
UreB (urease) protein	*Daucus carota*	Zhang et al. (2010)
Salmo salar (SasalFN-α1) protein	*S. tuberosum*	Fukuzawa et al. (2010)
	O. sativa	
GM-CSF	*N. tabacum*	Kim et al. (2004)
Human interleukin-10	*O. sativa* seeds	Fujiwara et al. (2010)

6.6.4 *Strategies for Increasing Yields*

Frequently, a drawback of plants as a platform for recombinant protein production is the low yields. Therefore, the challenge is to optimize the yield and quality of the protein produced. An exhaustive review can be found in Egelkrout et al (2012).

Briefly, the actions for increasing recombinant protein yields can be taken at three different stages:

6.6.4.1 Plant Cell Level

(a) transcriptional: by selecting promoters that best fit to our needs (strong, inducible), use of sequences that modulate gene expression (enhancer, activators, suppressors) (Dorokhov et al. 2002; Kasuga et al. 2004; Singer et al. 2010);
(b) translational: Optimization of 5′- and 3′- untranslated regions (UTR), codon design (e.g. codon vegetalization) (Sugio et al. 2008);
(c) post-translational: directing the expression and accumulation of the protein in a subcellular compartment with low levels of overall proteolytic activity (ER, chloroplasts, etc.) (Gomord et al. 1997; Ma et al. 2003; Chou and Shen 2010), directing the protein to the secretor pathway to collect it from the culture medium (e.g.: in *in vitro* cultures) (Hellwig et al. 2004; Komarnytsky et al. 2006; Lienard et al. 2007; Benchabane et al. 2008), co-expressing the protein with protease inhibitors (Faye et al. 2005; Michaud et al. 2005; Komarnytsky et al. 2006), or expressing the recombinant protein as a fusion protein with a stable and highly expressed peptide (Nygren et al. 1994; Stahl et al. 1997).

6.6.4.2 Optimization of the Environmental Conditions

The environmental conditions can be ameliorated by driving the accumulation of the recombinant protein to a determined more favorable organ (green or senescent leaves, seeds, tubers, roots), by the addition of protease inhibitors (EDTA, leupeptin, PMSF, etc.) to avoid the action of proteases present in the plant extract and by optimization of the culture conditions in *in vitro* cultures (Benchabane et al. 2008; Stoger et al. 2005; Michaud et al. 1998; Artsaenko et al. 1998). Another common strategy is the addition of protein stabilizing agents (gelatin, polyethylene glycol, polyvinylpyrrolidone, albumin) to the culture medium (Doran 2006b; Soderquist and Lee 2005).

– Addition of stabilizers: additives such as gelatin and polyvinylpyrrolidone (PVP) appear to be effective (Magnuson et al. 1996; Wongsamuth and Doran 1997; Doran 2006a etc.). The addition of PVP, human serum albumin showed certain limitations at upstream process level, such as foam formation (HAS), and at downstream process level, by a reduction of the binding efficiency on chromatography columns (PVP).

- Osmotic agents (mannitol, inositol), that inhibit cell disruption, have also influenced positively the accumulation of recombinant proteins (Lee et al. 2002; López et al. 2010).
- Coating agents (e.g.: Pluronic F127), the addition of such agents that avoid the adsorption of the recombinant protein to the vessels surfaces has demonstrated that increase the yield of some recombinant proteins (Doran 2006b).
- Oleosins are the main membrane proteins from oil bodies, they are alkaline and hydrophobic proteins with an amphipathic N- and C- termini, and a central hydrophobic domain. The hydrophobic domain is embedded in the phospholipid monolayer that surrounds the oil body, and the amphipathic domains are exposed to the surrounding environment (Bhatla et al. 2010). Oil bodies have the property of being easier to be separated from other cellular compartments and contaminating proteins by floating centrifugation (Ling 2007). The expression of recombinant proteins, fused to oleosines, in the surface of oil bodies is a strategy proposed to simplify the down stream processing. Some recombinant proteins have been expressed fused to oleosines such as hirudine, a specific thrombin inhibitor, that was expressed in *Brassica napus* and *B. carinata* (Parmenter et al. 1995; Chaudhary et al. 1998), a recombinant human precursor of insulin (Des-B30) and insulin were expressed in *Arabidopsis* (Nykiforuk et al. 2006; Markley et al. 2006), epidermal growth factor in *Arabidopsis* (Moloney and Van-Rooijen 2006), and also several antigenic peptides (Deckers et al. 2004).

6.6.4.3 Optimization of Large-Scale Processes

The selection of the bioreactor design is crucial (conventional, plastic-sleeves, etc.), as well as the mode of operation (batch, fed-batch, continuous), and the removal of the recombinant protein from the culture medium (if it is secreted) or from the biomass (if it is retained at subcellular level) (Doran 2000).

A remarkable aspect of conducting a productive process in a bioreactor is that it could be conducted following the regulatory requirements for the production of recombinant protein for therapeutics.

Down Stream Processing

When the protein is retained into the biomass, e.g., retained in the ER, the protein has to be extracted, a step that complicates the purification process. On the contrary, if the protein is targeted to the secretory pathway its recovery from the culture medium is easier and therefore less expensive (Baur et al. 2004). The stability of the recombinant protein also determines the strategy to be used, in some occasions the addition of stabilizers to the culture medium render highest yields (PVP, gelatin, etc.).

Also, the combination of ultrafiltration and chromatography was employed to purify a recombinant human monoclonal antibody (anti *Pseudomonas aeruginosa*

Table 6.4 Most common strategies employed to increase recombinant protein expression

Level		Strategy
Plant cell	Transcription	Strong, and/or inducible promoters
		Enhancers, activators and repressors
	Translation	Optimization of 5′- and 3′- UTR (untranslated regions)
		Codon design
	Post-translation	Driving to sub-cellular compartments, co-expression with protease inhibitors, cofactors. Expression as fusion proteins with a stable and highly expressed peptide
In vitro cultures		Optimization of the culture medium
		Addition of protein stabilizers
		In situ extraction of proteins
Bioreactor		Selection and improvement of bioreactor design
		Culture strategy (batch vs. fed-batch vs. continuous culture)
		Secretion into the culture medium

serotype O6ad immunoglobulin G1) with a high degree of removal of impurities and the complete recovery of the antibody. The authors included a two-stage cascade ultrafiltration process (a 50 kDa module and a 50 or 100 kDa MiniKros hollow fibre module, both operated at a shear rate of 1,800 s^{-1}) followed by a two-step chromatographic procedure (Mayani et al. 2013).

To facilitate the isolation and purification of the recombinant proteins by a Ni NTA affinity chromatography, a His6-tag sequence can be included in the construct.

Other strategy is to produce elastin-like polypeptides fusion proteins (ELPylation) to take advantage of the temperature-dependent, reversible aggregation/precipitation of ELPs and avoid tricky and expensive chromatographic steps, e.g.: for the expression of spider silk protein in the ER of tobacco and potato (Scheller et al. 2004; Floss et al. 2009).

Table 6.4 summarizes some of the most common strategies employed to improve plant-made protein yields.

6.7 Concluding Remarks

The production of recombinant proteins for medical purposes in plants is undoubtedly an alternative to the classical production plat-forms. However, their use is not yet commercially extended. The approval of taliglucerase alfa (ELELYSO®) is promissory but there are some drawbacks that have to be addressed.

Each recombinant protein is a particular challenge, thus generalization is not strictly possible making necessary to balance several aspects such as the plant species that will express the protein, if a the best strategy is a transient or a stable expression, if the protein will be expressed in the whole plant or in a particular organ or even in *in vitro* cultures, etc.

A considerable amount of functional proteins of animal origin have been expressed in plants including antibodies, antigens, enzymes, blood proteins, citoquines, growth factors and growth hormones (Davies 2010; De Muynck et al. 2010; Xu et al. 2012).

The plant-made proteins resulted functional in most of the cases because plants can perform the adequate folding, establishment of disulfide bonds, subunit assembly, proteolytic cut, and glycosylation.

Antibodies produced in plants were the focus of study for the significant cost reduction that could be attained in the case that the productions in plants were successful (De Muynck et al. 2010). In fact, it is considered that plants are the sole viable platform for producing recombinant secretory antibodies (IgAs) (Paul and Ma 2011).

References

Abranches R, Marcel S, Arcalis E, Altman F, Fevereiro P, Stoker E (2005) Plants as bioreactors: a comparative study suggests that *Medicago trunculata* is a promising production system. J Biotechnol 120:121–134

Ahmad P, Ashraf M, Younis M, Hu X, Kumar A, Akram NA, Al-Qurainy F (2012) Role of transgenic plants in agriculture and biopharming. Biotechnol Adv 30:524–540

An G (1985) High efficiency transformation of cultured tobacco cells. Plant Physiol 79:568–570

Artsaenko O, Ketting B, Fiedler U, Conrad U, Turing K (1998) Potato tubers as a biofactory for recombinant antibodies. Mol Breed 4:313–319

Austin S, Bingham ET, Mathews DE, Shahan MN, Will J, Burgess RR (1995) Production and field performance of transgenic alfalfa (*Medicago sativa* L.) expressing alpha-amylase and manganese-dependent lignin peroxidase. Euphytica 85:381–393

Awale MM, Mody SK, Dudharta GB, Avinash K, Patel HB et al (2012) Transgenic plant vaccine: a breakthrough in immunopharmacotherapeutics. J Vaccines 3(5). http://dx.doi.org/10.4172/2157-7560.1000147

Bardor M, Loutelier-Bourhis C, Paccalet T, Cosette P, Fitchette AC et al (2003) Monoclonal C5-1 antibody produced in transgenic alfalfa plants exhibits a N-glycosylation that is homogenous and suitable for glyco-engineering into human-compatible structures. Plant Biotechnol J 1:451–462

Bardor M, Cabrera G, Rudd PM, Dwek RA, Cremata JA, Lerouge P (2006) Analytical strategies to investigate plant N-glycan profiles in the context of plant-made pharmaceuticals. Curr Opin Struct Biol 16:576–583

Baur A, Reski R, Gorr G (2004) Enhanced recovery of a secreted recombinant human growth factor using stabilizing additives and by co-expression of human serum albumin in the moss *Physcomitrella patens*. Plant Biotechnol J 3(3):331–340

Benchabane M, Goulet C, Rivard D, Faye L, Gomord V, Michaud D (2008) Preventing unintended proteolysis in plant protein biofactories. Plant Biotechnol J 6:633–648

Bhatla SC, Kaushik V, Yadav MK (2010) Use of oil bodies and oleosins in recombinant protein production and other biotechnological applications. Biotechnol Adv 28:293–300

Bock R (2014) Genetic engineering of the chloroplast: novel tools and new applications. Curr Opin Biotechnol 26:7–13

Boothe J, Nykiforuk C, Shen Y, Zaplachinski S, Szarka S et al (2010) Seed-based expression systems for plant molecular farming. Plant Biotechnol J 8:588–606

Bosch D, Castillo A, Loos A, Schots A, Steinkellner H (2013) N-glycosylation of plant-produced recombinant proteins. Curr Pharm Des 19:1–10

Bouquin T, Thomson M, Nielsen LK, Green TH, Mundy J, Dziegiel MH (2002) Human anti-rhesus D IgG1 antibody produced in transgenic plants. Transgenic Res 11:115–122

Brereton HM, Chamberlain D, Yang R, Tea M, Mcneil S et al (2007) Single chain antibody fragments for ocular use produced at high levels in a commercial wheat variety. J Biotechnol 129:539–546

Brodzik R, Bandurska K, Deka D, Golovkin M, Koprowski H (2005) Advances in alfalfa mosaic virus-mediated expression of anthrax antigen in planta. Biochem Biophys Res Commun 338:717–722

Brodzik R, Spitsin S, Pogrebnyak N, Bandurska K, Portocarrero C, Andryszak Koprowski H, Golovkin M (2009) Generation of plant-derived recombinant DTP subunit vaccine. Vaccine 27:3730–3734

Buntru M, Gärtner S, Staib L, Kreuzaler F, Schalaich N (2013) Delivery of multiple transgenes to plant cells by an improved version of MultiRound Gateway technology. Transgenic Res 22: 153–167

Cañizares MC, Nicholson L, Lomonossoff GP (2005) Use of viral vectors for vaccine production in plants. Immunol Cell Biol 83:263–270

Cary JW, Rajasekaran K, Jaynes JM, Cleveland TE (2000) Transgenic expression of a gene encoding a synthetic antimicrobial peptide results in inhibition of fungal growth *in vitro* and in planta. Plant Sci 154:171–181

Castilho A, Gattinger P, Grass J, Jez J, Pabst M et al (2011) N-Glycosylation engineering of plants for the biosynthesis of glycoproteins with bisected and branched complex N-glycans. Glycobiology 21(6):813–823

Castilho A, Neumann L, Daskalova S, Mason HS, Steinkellner H, Altman F, Strasser R (2012) Engineering of sialylated mucin-type O-glycosylation in plants. J Biol Chem 287(43): 36518–36526

Cerovská N, Hoffmeisterová H, Pecenková T, Moravec T, Synková H, Plochová H, Velemínsky J (2008) Transient expression of HPV16 E7 peptide (aa 44–60) and HPV16 L2 peptide (aa 108–120) on chimeric potyvirus-like particles using Potato virus X-based vector. Protein Expr Purif 58:154–161

Chaudhary S, Parmenter DL, Moloney MM (1998) Transgenic *Brassica carinata* as a vehicle for the production of recombinant proteins in seeds. Plant Cell Rep 17:195–200

Chen QJ, Zhou HM, Chen J, Wang XC (2006) A gateway-based platform for multigene plant transformation. Plant Mol Biol 62:927–936

Cheo DL, Titus SA, Byrd DR, Hartley JL, Temple GF, Brasch MA (2004) Concerted assembly and cloning of multiple DNA segments using *in vitro* site-specific recombination. Functional analysis of multi-segment expression clones. Genome Res 14:2111–2120

Chikwamba RK, Scott MP, Mejia LB, Mason HS, Wang K (2003) Localization of a bacterial protein in starch granules of transgenic maize kernels. Proc Natl Acad Sci U S A 100: 11127–11132

Chou KC, Shen HB (2010) Plant-mPLoc: a top-down strategy to augment the power for predicting plant proteína subcellular localization. PLoS One 5:e11335

Chowdhury K, Bagasra O (2007) An edible vaccine for malaria using transgenic tomatoes of varying sizes, shapes and colors to carry different antigens. Med Hypotheses 68:22–30

Circelli P, Donini MM, Villani ME, Benvenuto E, Marusic C (2010) Efficient *Agrobacterium-based* transient expression system for the production of biopharmaceuticals in plants. Bioeng Bugs 1(3):221–224

Clarke JL, Daniell H, Nugent JM (2011) Chloroplast biotechnology, genomics and evolution: current status, challenges and future directions. Plant Mol Biol 76:207–209

Cramer CL, Weissenborn DL, Oishi KK, Grabau EA, Bennett S et al (1996) Bioproduction of human enzymes in transgenic tobacco. Ann NY Acad Sci 792:62–71

Cunha NB, Murad AM, Cipriano TM, Araújo ACG, Aragao FJL, Leite A et al (2011a) Expression of functional recombinant human growth hormone in transgenic soy-bean seeds. Transgenic Res 20:811–826

Cunha NB, Murad AM, Ramos GL, Maranhao AQ, Brígido MM et al (2011b) Accumulation of functional recombinant human coagulation factor IX in transgenic soybean seeds. Transgenic Res 20:841–855

D'Aoust MA, Lerouge P, Busse U, Bilodeau P, Trepanier S et al (2004) Efficient and reliable production of pharmaceuticals in alfalfa. In: Fischer R, Schillberg S (eds) Molecular farming. Wiley-VCH, Weinheim, pp 1–12

D'Aoust MA, Couture MJ, Charland N, Trépanier S, Landry N, Ors F, Vezina LP (2010) The production of hemagglutinin-based virus-like particles in plants: a rapid, efficient and safe response to pandemic influenza. Plant Biotechnol J 8:607–619

Daniell H, Streatfield SJ, Wycoff K (2001) Expression of the native cholera of toxin B subunit gene and assembly as functional oligomers in transgenic tobacco chloroplasts. J Mol Biol 311: 1001–1009

Daniell H, Chebolu S, Kumar S, Singleton M, Falconer R (2005) Chloroplast-derived vaccine antigens and other therapeutic proteins. Vaccine 23:1779–1783

Daniell H, Singh ND, Mason H, Streatfield J (2009) Plant-made vaccine antigens and biopharmaceuticals. Trends Plant Sci 14(12):669–679

Daskalova SM, Radder JE, Cichacz ZA, Olsen SH, Tsaprailis G, Mason H, López LC (2010) Engineering of N. benthamiana L. plants for production of N-acetylgalactosamine-glycosylated proteins – towards development of a plant-based platform for production of protein therapeutics with mucin type O-glycosylation. BMC Biotechnol 10:62, http://www.biomedcentral.com/1472-6750/10/62

Davies HM (2010) Commercialization of whole-plant systems for biomanufacturing of protein products: evolution and prospects. Plant Biotechnol J 8:845–861

Day A, Goldschmidt-Clermont M (2011) The chloroplast transformation toolbox: selectable markers and marker removal. Plant Biotechnol J 9:540–553. doi:10.1111/j.1467-7652.2011.00604.x

de la Riva GA, González-Cabrera J, Vázquez-Padrón R, Ayra-Pardo C (1998) Agrobacterium tumefaciens: a natural tool for plant transformation. Electron J Biotechnol 1(3):118–131

De Muynck B, Navarre C, Nizet Y, Stadlmann J, Boutry M (2009) Different subcellular localization and glycosylation for a functional antibody expressed *in Nicotiana tabacum* plants and suspension cells. Transgenic Res 18:467–482

De Muynck B, Navarre C, Boutry M et al (2010) Production of antibodies in plants: status after twenty years. Plant Biotechnol J 8:529–563

De Neve M, De Loose M, Jacobs A, Van Houdt H et al (1993) Assembly of an anti-body and its derived antibody fragment in *Nicotiana* and *Arabidopsis*. Transgenic Res 2:227–237

De Wilde C, Peeters K, Jacobs A, Peck I, Depicker A (2002) Expression of antibodies and Fab fragments in transgenic potato plants: a case study for bulk production in crop plants. Mol Breed 9:271–282

Deckers HM, Van Rooijen G, Boothe J, Goll J, Moloney MM, Schryvers AB et al (2004) Sembiosys Genetics Inc. Immunogenic formulations comprising oil bodies. United States patent US 6761914

Demain AL, Vaishnav P (2009) Production of recombinant proteins by microbes and higher organisms. Biotechnol Adv 27:297–306

Doran PM (2000) Foreign protein production in plant tissue cultures. Curr Opin Biotechnol 11:199–204

Doran PM (2006a) Foreign protein degradation and instability in plants and plant tissue cultures. Trends Biotechnol 24(9):426–432

Doran PM (2006b) Loss of secreted antibody from transgenic plant tissue cultures due to surface adsorption. J Biotechnol 122:39–54

Dorokhov YL, Skulachev MV, Ivanov PA, Zvereva SD, Tjulkina LG et al (2002) Polypurine (A)-rich sequences promote cross-kingdom conservation of internal ribosome entry. Proc Natl Acad Sci U S A 99:5301–5306

Dorokhov YL, Sheveleva AA, Frolova OY, Komarova TV, Zvereva AS et al (2007) Superexpression of tuberculosis antigens in plant leaves. Tuberculosis 87:218–224

Eeckhout D, De Clerq A, Van De Slijke E, Van Leene J, Stans H, Castells P et al (2004) A technology platform for the fast production of monoclonal recombinant antibodies against plant proteins and peptides. J Immunol Methods 294:181–187

Egelkrout E, Rajan V, Howard JA (2012) Overproduction of recombinant proteins in plants. Plant Sci 184:83–101

Faye L, Boulaflous A, Benchabane M, Gomord V, Michaud D (2005) Protein modifications in the plant secretory pathway: current status and practical implications in molecular pharming. Vaccine 23:1770–1778

Fernández-San Millán A, Mingo-Castel A, Miller M, Daniell H (2003) A chloroplast transgenic approach to hyper-express and purify human serum albumin, a protein highly susceptible to proteolytic degradation. Plant Biotechnol J 1:71–79

Fischer R, Twtman RM, Schillberg S (2003) Production of antibodies in plants and their use for global health. Vaccine 21:820–825

Fischer R, Stoger E, Schillberg S, Christou P, Twymann RM (2004) Plant-based production of biopharmaceuticals. Curr Opin Plant Biol 7:152–158

Floss DM, Schallau KM, Rose-John S, Conrad U, Scheller J (2009) Elastin-like polypeptides revolutionize recombinant protein expression and their biomedical application. Trends Biotechnol 28(1):37–45

Franconi R, Demurtas OC, Massa S (2010) Plant-derived vaccines and other therapeutics produced in contained systems. Expert Rev Vaccines 9(8):877–892

Fujiwara Y, Aiki Y, Yang L, Takaiwa F, Kosaka A, Tsuji NM, Shiraki K, Sekikawa K (2010) Extraction and purification of human interleukin-10 from transgenic rice seeds. Protein Expr Purif 72:125–130

Fukuzawa N, Tabayashi N, Okinaka Y, Furusawa R, Furuta K, Kagaya U, Matsumura T (2010) Production of biologically active Atlantic salmon interferon in transgenic potato and rice plants. J Biosci Bioeng 110:201–207

Galeffi P, Lombardi A, Di Donato M, Latini A, Sperandei M, Cantale C, Giacomini P (2005) Expression of single-chain antibodies in transgenic plants. Vaccine 23:1823–1827

Gao Y, Ma Y, Li M, Cheng T, Li SW, Zhang J, Xia NS (2003) Oral immunization of animals with transgenic cherry tomatillo expressing HBsAg. World J Gastroenterol 9:996–1002

Gao J, Hooker BS, Anderson DB (2004) Expression of functional human coagulation factor XIII A-domain in plant cell suspensions and whole plants. Protein Expr Purif 37:89–96

Gao M, Li Y, Xue X, Wang X, Long J (2012) Stable plastid transformation for high-level recombinant protein expression: promises and challenges. J Biomed Biotechnol, Article ID 158232, 16 pp. doi:10.1155/2012/158232

Gelvin SB (2003) Agrobacterium-mediated plant transformation: the Biology behind the "Gene-Jockeying" tool. Microbiol Mol Biol Rev 67(1):16–37

Giorgi C, Franconi R, Rybicki E (2010) Human papillomavirus vaccines in plants. Expert Rev Vaccines 9(8):913–924

Gleba Y, Klimyuk V, Marillonet S (2005) Magnifection – a new platform for expressing recombinant vaccines in plants. Vaccine 23:2042–2048

Gomord V, Faye L (2004) Posttranslational modification of therapeutic proteins in plants. Curr Opin Plant Biol 7:171–181

Gomord V, Denmat LA, Fitchette AC, Satiat-Jeunemaître B, Hawes C, Faye L (1997) The C-terminal HDEL sequence is sufficient for retention of secretory proteins in the endoplasmic reticulum (ER) but promotes vacuolar targeting of proteins that escape the ER. Plant J 11:313–325

Gomord V, Fitchette AC, Menu-Bouaouiche L, Saint-Jore-Dupas C, Plasson C, Michaud D, Faye L (2010) Plant-specific glycosylation patterns in the context of therapeutic protein production. Plant Biotechnol J 8:564–587

Haq TA, Mason HS, Clements JD, Arntzen CJ (1995) Oral immunization with a recombinant bacterial antigen produced in transgenic plants. Science 268:714–716

Hartley JL, Temple GF, Brasch MA (2000) DNA cloning using *in vitro* site-specific re-combination. Genome Res 10:1788–1795

Hellens RP, Edwards EA, Lelyland NR, Bean S, Mullineaux PM (2000) pGREEN: a versatile and flexible binary Ti vector for *Agrobacterium*-mediated plant trans-formation. Plant Mol Biol 42:819–832

Hellwig S, Drossard J, Twyman R, Fischer R (2004) Plant cell cultures for the production of recombinant proteins. Nat Biotechnol 22(11):1414–1422

Herrera-Estrella L, Depicker A, Van Montagu M, Scholl J (1983) Expression of chimeric genes transferred into plant cells using a Ti-plasmid-derived vector. Nature 303:209–213

Hiatt A, Cafferkey R, Bowdish K (1989) Production of antibodies in transgenic plants. Nature 342:76–78

Hong SH, Kim KI, Chung HY, Kim YJ, Sunter G, Bisaro DM, Chung IS (2004) Expression of recombinant endostatin in *Agrobacterium*-inoculated leaf disks of *Nicotiana tabacum* var. Xanthi. Biotechnol Lett 26:1433–1439

Hood EE (2002) From green plants to industrial enzymes. Enzyme Microb Technol 30:279–283

Hood EE, Woodard SL, Horn ME (2002) Monoclonal antibody manufacturing in transgenic plants – myths and realities. Curr Opin Biotechnol 13:630–635

Huang Z, Mason HS (2004) Conformational analysis of hepatitis B surface antigen fusions in an Agrobacterium-mediated transient expression system. Plant Biotechnol J 2:241–249

James E, Mills DR, Lee JM (2002) Increased production and recovery of secreted foreign proteins from plant cell cultures using an affinity chromatography bioreactor. Biochem Eng J 12:205–213

Jeffrey MS, Garcia B, Graves J, Hajdukiewicz PTJ, Hunter P et al (2000) High-yield production of a human therapeutic protein in tobacco chloroplasts. Nat Biotechnol 18:333–338

Joshi L, López LC (2005) Bioprospecting in plants for engineered proteins. Curr Opin Plant Biol 8:223–226

Kaldis A, Ahmad A, Reid A, Mcgarvey B, Brandle J, Ma S et al (2013) High-level production of human interleukin-10 fusions in tobacco cell suspension cultures. Plant Biotechnol J 11: 535–545

Kapila J, De Rycke R, Van Montagu M, Angenon G (1997) An Agrobacterium-mediated transient gene expression system for intact leaves. Plant Sci 122:101–108

Karimi J, Inzé D, Depicker A (2002) GATEWAY® vectors for *Agrobacterium*-mediated plant transformation. Trends Plant Sci 7(5):193–195

Karimi M, De Meyer B, Hilson P (2005) Modular cloning in plant cells. Trends Plant Sci 10:103–105

Karimi M, Depicker A, Hilson P (2007) Recombinational cloning with plant gateway-vectors. Plant Physiol 145(4):1144–1154

Kasuga M, Miura S, Shinozaki K, Yamaguchi-Shinozaki K (2004) A combination of the *Arabidopsis* DREB1A gene and stress-inducible rd29A promoter improved drought-and low-temperature stress tolerance in tobacco by gene transfer. Plant Cell Physiol 45:346–350

Kim YS, Kwon TH, Sik YM (2004) Direct transfer and expression of human GM-CSF in tobacco suspension cell using *Agrobacterium*-mediated transfer system. Plant Cell Tiss Org Cult 78:133–138

Komarnytsky S, Borisjuk N, Yakoby N, Garvey A, Raskin I (2006) Co-secretion of protease inhibitor stabilizes antibodies produced by plant roots. Plant Physiol 141:1185–1193

Kong Q, Richter L, Yang YF, Arntzen CJ, Mason HS, Thanavala Y (2001) Oral immunization with hepatitis B surface antigen expressed in transgenic plants. Proc Natl Acad Sci U S A 98:11539–11544

Koprivova A, Stemmer C, Altman F, Hoffmann A, Kopriva S, Gorr G, Reiki R, Decker EL (2004) Targeted knockouts of *Physcomitrella* lacking plant-specific immunogenics N-glycans. Plant Biotechnol J 2:517–523

Kostrzak A, Gonzalez MC, Guetard D, Nagaraju DB, Hobson SW, Tepfer D, Pniewski T, Ala M (2009) Oral administration of low doses of plant based HBsAg induced antigen specific IgAs and IgGs in mice, without increasing levels of regulatory T cells. Vaccine 27:4798–4807

Landry N, Trépanier S, Montomoli E, Dargis M, Papini G, Vézina LP (2010) Preclinical and clinical development of plant-made virus-like particle vaccine against avian H5N1 influenza. PLoS One 5(12):e15559. doi:10.1371/journal.pone.0015559

Larrick JW, Yu L, Naftzger C, Jaiswal S, Wycoff K (2001) Production of secretory IgA antibodies in plants. Biomol Eng 18:87–94

Lee JH, Kim NS, Kwon TH, Yang MS (2002) Effects of osmotic pressure on production of recombinant human granulocyte-macrophage colony stimulating factor in plant cell suspension culture. Enzyme Microb Technol 30:768–773

Lei H, Qi J, Song J, Yang D, Wang Y, Zhang Y, Yang J (2006) Biosynthesis and bio-activity of trichosanthin in cultured crown gall Maximowicz. Plant Cell Rep 25:1205–1212

Lerouge P, Nardor M, Lagny S, Gomord V, Faye L (2000) N-glycosylation of recombinant pharmaceutical glycoproteins produced in glycoproteins produced in transgenic plants: towards an humanisation of plant N-glycans. Curr Pharm Biotechnol 1:347–354

Lienard D, Tran Dinh O, Van Oort E, Van Overtvelt L, Bonneau C et al (2007) Suspension-cultured BY-2 tobacco cells produce and mature immunologically active house dust mite allergens. Plant Biotechnol J 5:93–108

Ling H (2007) Oleosin fusion expression systems for the production of recombinant proteins. Biologia 62:119–123

López J, Lencina F, Petruccelli S, Marconi P, Alvarez MA (2010) Influence of the KDEL signal, DMSO and mannitol on the production of the recombinant antibody 14D9 by long-term Nicotiana tabacum cell suspension culture. Plant Cell Tiss Org Cult 103:307–314

Ma JKC, Hein M (1995) Immunotherapeutic potential of antibodies produced in plants. Trends Biotechnol 13:522–527

Ma KC, Hikmat BY, Wicoff K, Vine ND, Chargelegue D et al (1998) Characterization of a recombinant plant monoclonal antibody and preventive immunotherapy in humans. Nat Med 4(5): 601–606

Ma JK, Drake PM, Christou P (2003) The production of recombinant pharmaceutical proteins in plants. Nat Rev Genet 4:794–805

Magnani E, Bartling L, Hake S (2006) From gateway to MultiSite gateway in one re-combination event. BMC Mol Biol 7:46

Magnuson NS, Linzmaier PM, Gao JW, Reeves R, An G, Lee JM (1996) Enhanced recovery of secreted mammalian protein from suspension culture of genetically modified tobacco cells. Protein Expr Purif 7:220–228

Magnuson NS, Linzmaier PM, Reeves R, An G, Hayglass K, Lee JM (1998) Secretion of biologically active human interleukin-2 and interleukin-4 from genetically modified tobacco cells in suspension culture. Protein Expr Purif 13:45–52

Marconi PL, Alvarez MA (2014) State of the art on plant-made single-domain antibodies. J Immunol Tech Infect Dis 3:2. doi: 10.4172/2329-9541.1000125

Marillonnet S, Giritch A, Gils M, Kandzia R, Klimyuk V, Gleba Y (2004) In planta engineering of viral RNA replicons: efficient assembly by recombination of DNA modules delivered by Agrobacterium. Proc Natl Acad Sci U S A 101:5852–5857

Markley N, Nykiforuk C, Moloney MM (2006) Producing proteins using transgenic oilbody-oleosin technology. Biopharm Int 19:6

Martínez-González L, Rosales-Mendoza S, Soria-Guerra RE, Moreno-Fierros L, López-Reviña R et al (2011) Oral immunization with a lettuce-derived *Escherichia coli* heat-labile toxin B subunit induces neutralizing antibodies in mice. Plant Cell Tiss Org Cult 107:441–449

Mascia PN, Flavell RB (2004) Safe and acceptable strategies for producing foreign molecules in plants. Curr Opin Plant Biol 7:189–195

Mason HS, Arntzen CJ (1995) Transgenic plants as vaccine production systems. Trends Biotechnol 13(9):388–392

Mason HS, Warzecha H, Mor T, Arntzen CJ (2002) Edible plant vaccines: applications for prophylactic and therapeutic molecular medicine. Trends Mol Med 8(7):324–329

Matsuda R, Kubota C, Alvarez ML, Cardineau GA (2009) Biopharmaceutical protein production under controlled environments: growth, development, and vaccine productivity of transgenic tomato plants grown hydroponically in a greenhouse. HortScience 44(6):1594–1599

Matsumoto Y, Suzuki S, Nozoye T, Yamakawa T, Takashima Y et al (2009) Oral immunogenicity and protective efficacy in mice of transgenic rice plants producing a vaccine candidate antigen (As16) of Ascaris suum fused with cholera toxin B subunit. Transgenic Res 18:185–192

Mayani M, Filipe CDM, Mclean MD, Hall JC, Ghosh R (2013) Purification of transgenic tobacco-derived recombinant human monoclonal antibody. Biochem Eng J 72:33–41

Mcdonald KA, Hong LM, Trombly DM, Xie Q, Jackman AP (2005) Production of human alpha 1-antitrypsin from transgenic rice cell culture in a membrane bioreactor. Biotechnol Prog 21:728–734

Medina-Bolivar F, Cramer C (2004) Production of recombinant proteins by hairy roots cultured in plastic sleeve bioreactors. Methods Mol Biol 267:351–363

Medina-Bolivar F, Wright R, Funk V, Sentz D, Barroso L et al (2003) A non-toxic lectin for antigen delivery of plant-based mucosal vaccines. Vaccine 21(9):997–1005

Medrano G, Reidy MJ, Liu J, Ayala J, Dolan MC, Cramer CL (2009) Rapid system for evaluating bioproduction capacity of complex pharmaceutical proteins in plants. Methods Mol Biol 483:51–67

Menassa R, Nguyen V, Jevnikar A, Brandle J (2001) A self-contained system for the field production of plant recombinant interleukin-10. Mol Breed 8:177–185

Michaud D, Vrain TC, Gomord V, Faye L (1998) Stability of recombinant proteins in plants. Methods Biotechnol 3:177–188

Michaud D, Anguenot R, Brunelle F (2005) Method for increasing protein content in plant cells. US Patent application 2005/0055746

Moloney MM, Van-Rooijen G (2006) Sembiosys Genetics Inc. Expression of epidermal growth factor in plant seeds. United States patent US 7332587

Nygren PA, Stahl S, Uhlén M (1994) Engineering proteins to facilitate bioprocessing. Trends Biotechnol 12:181–188

Nykiforuk CL, Boothe JG, Murray EW, Keon RG, Goren J, Markley NA et al (2006) Transgenic expression and recovery of biologically active recombinant human insulin from *Arabidopsis thaliana* seeds. Plant Biotechnol J 4:77–85

Obembe OO, Popoola JO, Leelavathi S, Reddy SV (2011) Advances in plant molecular farming. Biotechnol Adv 29:210–222

Parmenter DL, Boothe JG, Van Rooijen GJH, Yeung EC, Moloney MM (1995) Production of biologically active hirudin in plants seeds using oleosin partitioning. Plant Mol Biol 29: 1167–1180

Paul M, Ma JKC (2011) Plant-made pharmaceuticals: leading products and production plat-forms. Biotechnol Appl Biochem 58:58–67

Peter RE, Yu KL, Marchant TA, Rosenblum PM (1990) Direct neural regulation of the teleost adenohypophysis. J Exp Zool 4:84–89

Raju TS, Briggs JB, Chamow SM, Winkler ME, Jones AJ (2001) Glycoengineering of therapeutic glycoproteins: *in vitro* galactosylation and sialylation of glycoproteins with terminal N-acetylglucosamine and galactose residues. Biochemistry 40:8868–8876

Ramessar K, Sabalza M, Capell T, Christou P (2008) Maize plants: an ideal production platform for effective and safe molecular pharming. Plant Sci 174:409–419

Ramessar K, Rademachar T, Sack M, Stadlmann J, Platis D et al (2009) Cost-effective production of a vaginal protein microbicide to prevent HIV transmission. Proc Natl Acad Sci USA 105(10):3727–3732

Sainsbury F, Lomonossoff GP (2008) Extremely high-level and rapid transient protein production in plants without the use of viral replication. Plant Physiol 148:1212–1218

Scheller J, Henggeler D, Viviani A, Conrad U (2004) Purification of spider silk-elastin from transgenic plants and application for human chondrocyte proliferation. Transgenic Res 13:51–57

Seddas-Dozolme P, Walter B, Van Regenmortel MHV (1999) Introduction: antigens, antibodies and plantibodies. In: Harper K, Ziegler A (eds) Recombinant anti-bodies. Applications in plant science and plant pathology. Taylor & Francis, London, pp 3–21

Sharma AK, Sharma MK (2009) Plants as bioreactors: recent developments and emerging opportunities. Biotechnol Adv 27:811–832

Sheludko YV (2008) Agrobacterium-mediated transient expression as an approach to production of recombinant proteins in plants. Recent Pat Biotechnol 2:198–208

Singer SD, Hily JM, Liu Z (2010) A1-kb Bacteriophage lambda fragment functions as an insulator to effectively block enhancer–promoter interactions in *Arabidopsis thaliana*. Plant Mol Biol Rep 28(1):69–76

Soderquist RG, Lee JM (2005) Enhanced production of recombinant proteins from plant cells by the application of osmotic stress and protein stabilization. Plant Cell Rep 24:127–132

Stahl S, Nygren PA, Uhlén M (1997) Detection and isolation of recombinant proteins based on binding affinity of reporter: protein A. Methods Mol Biol 63:103–118

Staub JM, García B, Graves J, Hajdukiewicz PT, Hunter P et al (2000) High-yield production of a human therapeutic protein in tobacco chloroplasts. Nat Biotechnol 18(3):333–338

Stoger E, Ma JK, Fischer R, Christou P (2005) Sowing the seeds of success: pharmaceutical proteins from plants. Curr Opin Biotechnol 16:167–173

Streatfield SJ, Howard JA (2003) Plant-based vaccines. Int J Parasitol 33:479–493

Sugio T, Satoh J, Matsuura H, Shinmyo A, Ato K (2008) The 5′-untranslated region of the *Oryza sativa* alcohol dehydrogenase gene functions as a translational enhancer in monocotyledonous plant cells. J Biosci Bioeng 105:300–302

Tanaka Y, Kimura T, Hikino K, Goto S, Nishimura M et al (2012) Gateway vectors for plant genetic engineering: overview of plant vectors, application for Bimolecular Fluorescence Complementation (BiFC) and multigene construction. In: Barrera-Saldaña HA (ed) Genetic engineering – basics, new applications and responsibilities. InTech Open Acces, Croatia, pp 35–58

Terashima M, Murai Y, Kawamura M, Nakanishi S, Stoltz T, Chen L, Drohan W, Rodríguez RL, Katoh S (1999) Production of human α1-antitrypsin by plant cell culture. Appl Microbiol Biotechnol 52:516–523

Thanavala Y, Mahoney M, Pal S, Scott A, Richter L et al (2005) Immunogenicity in humans of an edible vaccine for hepatitis B. Proc Natl Acad Sci U S A 102:3378–3382

Thomas BR, Van Deynze A, Bradford KJ (2002) Production of therapeutic proteins in plants. Agricultural Biotechnology in California Series, University of California, Agriculture and Natural Resources, Oakland, pp 1–12

Thyssen G, Svab Z, Maliga P (2012) Exceptional inheritance of plastids via pollen in *Nicotiana sylvestris* with no detectable paternal mitochondrial DNA in the progeny. Plant J 72:84–88

Tregoning J, Maliga P, Dougan G, Nixon PJ (2004) New advances in the production of edible plant vaccines: chloroplast expression of a tetanus vaccine antigen, TetC. Phytochemistry 65: 989–994

Tremblay R, Wang D, Jevnikar AM, Ma S (2010) Tobacco, a highly efficient green bioreactor for production of therapeutic proteins. Biotechnol Adv 28:214–221

Twymann RM, Soger E, Schillbeg S, Christou P, Fischer R (2003) Molecular farming in plants: host systems and expression technology. Trends Biotechnol 21(12):570–578

Verch T, Yusibov V, Koprowski H (1998) Expression and assembly of a full-length monoclonal antibody in plants using a plant virus vector. J Immunol Methods 220:69–75

Walmsley AM, Arntzen CJ (2003) Plant cell factories and mucosal vaccines. Curr Opin Biotechnol 14:145–150

Watson J, Koya V, Leppla SH, Daniell H (2004) Expression of *Bacillus anthracis* protective antigen in transgenic chloroplasts of tobacco, a non-food/feed crop. Vaccine 22:4374–4384

Weathers PJ, Giles KL (1988) Regeneration of plants using nutrient mists. In Vitro Cell Dev Biol Plant 24:727–732

Whaley KJ, Hiatt A, Zeltlin L (2011) Emerging antibody products and *Nicotiana* manufacturing. Hum Vaccin 7(3):349–356

Wongsamuth R, Doran PM (1997) Production of monoclonal antibodies by tobacco hairy roots. Biotechnol Bioeng 54:401–415

Woodard SL, Mayor JM, Bailey MR, Barker DK, Love RT et al (2003) Maize (*Zea mays*) derived bovine trypsin: characterization of the first large scale, commercial protein product from transgenic plants. Biotechnol Appl Biochem 38:123–130

Woods RR, Geyer BC, Mor TS (2008) Hairy-root organ cultures for the production of human acetylcholinesterase. BMC Biotechnol 8:95. doi:10.1186/1472-6750-8-95

Xu J, Dolan MC, Medrano G, Carol CL, Weathers PJ (2012) Green factory: plants as bioproduction platforms for recombinant proteins. Biotechnol Adv 30:1171–1184

Yusibov VM, Mamedov TG (2010) Plants as an alternative system for expression of vaccine antigens. Proc ANAS (Biol Sci) 65(5–6):195–200

Zhang H, Liu M, Li Y, Zhao Y, He H, Yang G, Zheng C (2010) Oral immunogenicity and protective efficacy in mice of a carrot-derived vaccine candidate expressing UreB subunit against *Helicobacter pylori*. Protein Expr Purif 69:127–131

Chapter 7
The Antibody 14D9 as an Experimental Model for Molecular Farming

Abstract The catalytic antibody 14D9 is an IgG1-like murine antibody that catalyzes the highly enantioselective protonation of enol-ethers. It was utilised as experimental model for studying the expression of an antibody in tobacco.

The ability of *Nicotiana tabacum* to express the whole antibody was confirmed. Also, *in vitro* cultures of *N. tabacum* were established and their ability to express the antibody was demonstrated. Attempts to improve the 14D9 initial yields were performed by the addition of a KDEL endoplasmic reticulum (ER) retention signal to the construct, therefore two *N. tabacum* lines were used in all the experiments, one with the secretory variant of the antibody (sec-Ab), and another one that retains the antibody in the ER (Ab-KDEL).

Other strategies tested were the optimization of plant growth regulator balance in the culture medium, and the addition of protein stabilizers, plant cell wall permeabilizers and osmotic agents. Besides, hairy root cultures were established and the conditions to express 14D9 were analyzed.

Comparing the yields obtained in all the platforms and system examined, we can conclude that hairy roots growing in Erlenmeyer flasks and Ab-KDEL cell suspension cultures growing in a 2-l bioreactor gave the highest 14D9 yields.

Keywords Catalytic antibody • 14D9 • *Nicotiana tabacum*-recombinant antibodies • Cell suspension cultures • Hairy roots

7.1 Introduction

Antibodies were among the first heterologous proteins expressed in plants (Hiatt et al. 1989). Soon after plants were engineered to express antibody fragments such as Fab, Fv, scFv, VHH, etc. (Desai et al. 2010; Egelkrout et al. 2012; Marconi and Alvarez 2014a).

As discussed in previous chapters, the production of recombinant proteins in *in vitro* cultures has additional advantages over whole plants. *In vitro*, the whole process can be performed in a shortest period of time, in environmental controlled conditions and good manufacturing and good laboratory practices (GMP and GLP) as is required by the pharmaceutical industry (Hiatt et al. 1989; Horsch et al. 1984; Paul and Ma 2011).

M.A. Alvarez, *Plant Biotechnology for Health: From Secondary Metabolites to Molecular Farming*, DOI 10.1007/978-3-319-05771-2_7

Among the antibodies, the catalytic antibodies are of particular interest for their capability to catalyse chemical reactions with a high regio- and stereoselectivity (Sinha et al. 1993). Furthermore, they can catalyse chemical transformations on demand, even in the case of reactions where no natural enzyme are available, making their production attractive to the chemical industry (Tanaka 2002).

In general, catalytic antibodies are produced by chemical synthesis or semisynthesis (Hasserodt 1999; Reymond 1999). They can also be expressed as recombinant antibodies in bacteria (Zheng et al. 2003), yeasts (Ulrich et al. 1995), or plants (Petruccelli et al. 2006). As for the applicability of those antibodies, they can be used in human and animal health, in diagnosis kits, and for organic synthesis.

7.2 The Antibody 14D9

The antibody 14D9 is an IgG_1-type murine antibody obtained by immunization against a piperidinium hapten. It is one of the fastest and most practical catalytic antibodies to date, being applicable for an enantioselective gram-scale synthesis (Reymond et al. 1991; Janghiri and Reymond 1994; Reymond et al. 1994; Zheng et al. 2004). 14D9, mechanistically related to enzymes as glycosidases, catalyses the highly enantioselective (>99 % ee) protonation of enol-ethers (Reymond 1999). It hydrolyzes a relatively broad spectrum of compounds with an aromatic nucleus, which is the recognition element. The best substrates for 14D9 are those with a stereochemistry (Z) in the double bond and an alquil substituent in the carbon to be protonated (Reymond et al 1993; Zheng et al. 2003).

As recombinant protein, 14D9 was expressed as a single-chain fragment (scFv), fused to the N utilization substance protein A (scFv-NusA), and as a chimeric fragment (Fab) in *E. coli* (Zheng et al. 2003). In *Nicotiana tabacum* plants 14D9 was expressed as the whole antibody (Petruccelli et al. 2006).

The potential use of 14D9 for releasing an essential growth factor (e.g.: biotin) from a specifically synthesized substrate is under study (Zheng et al. 2003).

7.2.1 14D9-Antibody Extraction and Analysis

For antibody extraction, samples are grinded using a cold mortar and pestle in cold PBS (0.24 g l^{-1} KH_2PO_4, 1.44 g l^{-1} Na_2HPO_4, 0.2 g l^{-1} KCl, 8 g l^{-1} NaCl, pH: 7.0–7.2) containing 10 μg ml^{-1} leupeptin. The extracts are then centrifuged at 14,000 g for 20 min at 4 °C and the supernatant is separated to measure the antibody. Concentrations of antibodies possessing both heavy (γ) and light (κ) chains are evaluated by for γ and κ chains, a mouse IgG is used as a standard (Sigma Chemical Co). Only antibodies sandwich enzyme-linked immunosorbent assay (ELISA) using goat anti-mouse antibodies specific assembled into γ –κ chain complexes are detected. The functionality of the antibody is verified by competitive ELISA using bovine serum albumin (BSA) coupled to the 14D9 haptene (Petruccelli et al. 2006).

7.3 Establishment of Hairy Roots Expressing the Antibody 14D9

Transgenic tobacco expressing 14D9 were obtained by transformation with *Agrobacterium tumefaciens* LBA4404 (Petruccelli et al. 2006). Briefly, the gamma (γ) and kappa (κ) chains codifying for 14D9 were isolated from the mRNA of hybridoma cells and cloned into the binary plasmid pGA482. The 35S promoter of Cauliflower mosaic virus (CaM35S), the Tobacco-etch virus (TEV) leader signal, and the CaM35S terminator were also in the construct. A second version was designed, bearing the endoplasmic reticulum (ER) retention signal KDEL at the COOH terminal of the κ chain. The plasmids were introduced by electroporation in *Agrobacterium* spp. After stable transformation of tobacco, several strains expressing the chains γ, κ o κ-KDEL were obtained, a version expressing γ and κ chains (sec-Ab) and another expressing the γ and κ –KDEL chains (Ab-KDEL).

Seeds of both versions of recombinant plants, sec-Ab and Ab-KDEL, were germinated *in vitro*. Leaves and stems of plantlets were infected with *A. rhizogenes* LBA 9402 (Martinez et al. 2005). After 15 days of infection, roots tips emerged at the inoculation points (Fig. 7.1). The frequency of transformation was 98 % for the sec-Ab version and 87.5 % for the Ab-KDEL version. The nascent roots were transferred to Petri dishes with MSRT medium plus kanamycin (100 μg ml^{-1}) to eliminate *Agrobacterium* (Fig. 7.2). Transformation was confirmed by PCR. The hairy roots lines obtained (sec-Ab and Ab-KDEL) have different patterns of growth and 14D9 expression. The best producing lines of each version were selected and used in the following experiments. Wild type hairy roots were used as control.

After *Agrobacterium* was eliminated, hairy roots (sec-Ab, Ab-KDEL and wild type) were transferred to MSRT liquid medium without plant growth regulators (PGR) or antibiotics and maintained in culture by periodical transferences to identical fresh media (Alvarez et al. 1994). After 6 months in culture, hairy roots

Fig. 7.1 Hairy root tips appearing in the infection sites 15 days post-infection

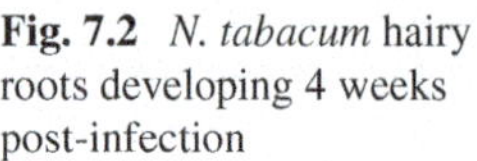

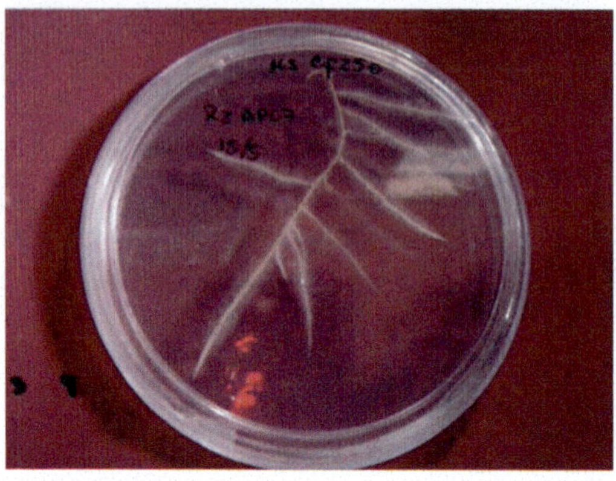

Fig. 7.2 *N. tabacum* hairy roots developing 4 weeks post-infection

were used to analyze growth and 14D9 production (Fig. 7.3). Hairy roots were transferred to a 2-bioreactor (Fig. 7.8), further experiments are being conducted in order to establish parameters of growth and 14D9 expression.

7.3.1 Kinetic of Growth and 14D9 Expression

Hairy roots of the sec-Ab, Ab-KDEL or wild type lines (% % W/V inoculum) were transferred to Erlenmeyer flasks containing MSRT culture medium. They were maintained at 24 ± 2 °C, a 16-hs photoperiod given by fluorescent lamps (irradiance 1.8 w m^{-1} s^{-1}) in a rotary shaker at 100 rpm during 25 days. Samples were taken by triplicate every five days. Cell growth (as DW) and 14D9 content (Martínez et al. 2005) were determined in each sample (Table 7.1). The specific growth rate showed significant differences ($P < 0.01$) between the sec-Ab and Ab-KDEL lines respect to the wild type. The growth of wild type hairy roots was characterized by a continuous increase of biomass during the first 10 days followed by a 15 days-stationary phase. On the other hand, the exponential growth phase of the sec-Ab line extended up to the 20th day of culture, whereas in the case of the Ab-KDEL line the exponential phase continues up to the 25th day of culture (Fig. 7.4). Growth Index (GI) was higher for the Ab-KDEL line than for the sec-Ab line. The maximal biomass in the Ab-KDEL line was higher than in sec-Ab line. The antibody amount in the biomass in the sec-Ab line was lower than in the Ab-KDEL line, 14D9 was not detected in the culture medium (Martínez et al. 2005). The presence of the KDEL sequence increases the level of recombinant protein delivered into the secretory pathway, as happened with other recombinant proteins (Yoshida et al. 2004).

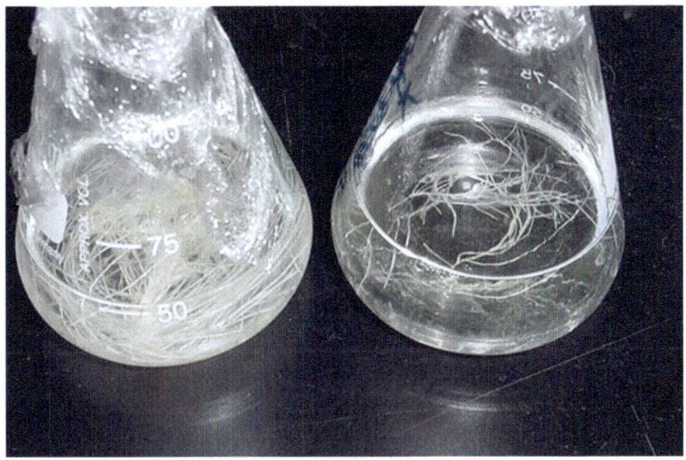

Fig. 7.3 *N. tabacum* hairy roots growing in MSRT culture medium without the addition of plant growth regulators

Table 7.1 Parameters of growth and 14D9 concentration in the biomass of *N. tabacum* wild type, sec-Ab and Ab-KDEL hairy root lines

N. tabacum line	μ (day^{-1})	GI	μ g14D9 mg FW^{-1}
Wild type	0.11 ± 0.0.5	0.98 ± 0.09	–
Sec-Ab	0.06 ± 0.01	1.03 ± 0.11	0.15 ± 0.01
Ab-KDEL	0.19 ± 0.09	6.45 ± 0.61	9.1 ± 0.09

7.3.2 *Influence of the Additives Dimethyl Sulfoxide (DMSO), Gelatine and Polyvinylpyrrolidone (PVP) on 14D9 Expression*

In order to increase 14D9 yields, the effect of dimethylsulfoxide (DMSO), gelatine and polyvinylpyrrolidone (PVP) was tested. Gelatine and PVP were selected for their activity as protein stabilizers, DMSO for the combined stabilizing and permeabilizing effect (Wongsamuth and Doran 1997).

7.3.2.1 Influence of Dimethylsulfoxide (DMSO)

An inoculum of 200 mg FW of hairy roots was transferred to 150 ml Erlenmeyer flasks containing 40 ml of MSRT medium without PGR. DMSO was added to the culture medium at different concentrations (2.8, 5.0, 8.0, and 10.0 % V/V) at the

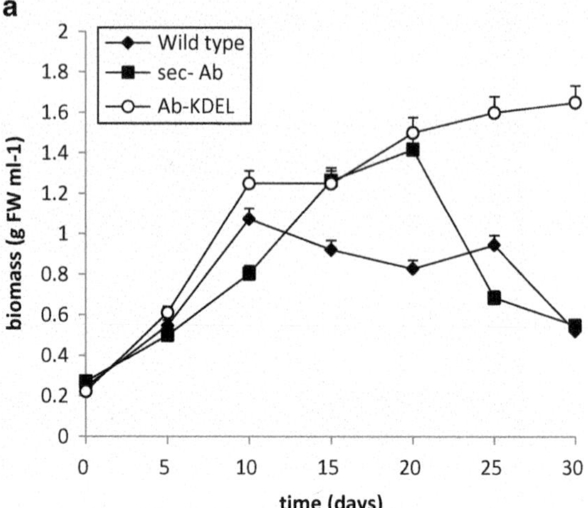

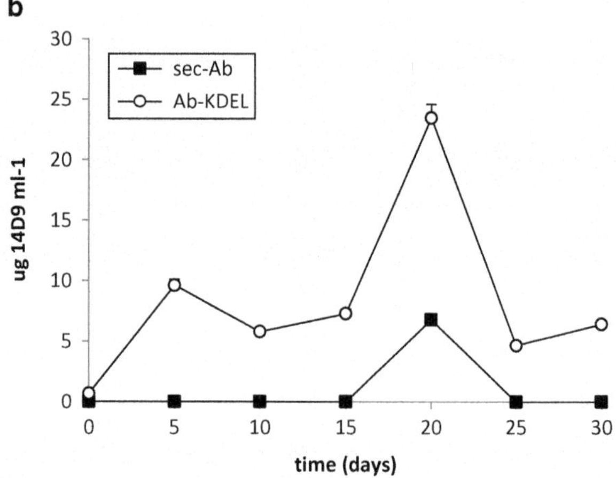

Fig. 7.4 Profile of (**a**) growth of sec-Ab, Ab-KDEL and wild type *N. tabacum* hairy roots, and (**b**) 14D9 expression in sec-Ab and Ab-KDEL hairy root lines

14th day of culture (Fig. 7.5). DMSO had a deleterious effect on hairy roots growth at all the concentrations tested. As for 14D9 content, DMSO has a positive effect on the amount of 14D9 in the biomass of both transgenic lines (Fig. 7.6a), and on the protein total content (Fig. 7.6b) in the Ab-KDEL line. Those results could be a consequence of the permeabilising effect of DMSO facilitating the entry of nutrients in the cell (e.g.: aminoacids) to be used for protein synthesis (Whal et al. 1995). Also, it was postulated that DMSO could be acting as a chemical chaperone that stabilizes proteins in their native conformation with a protective effect against proteases (Brown et al. 1996).

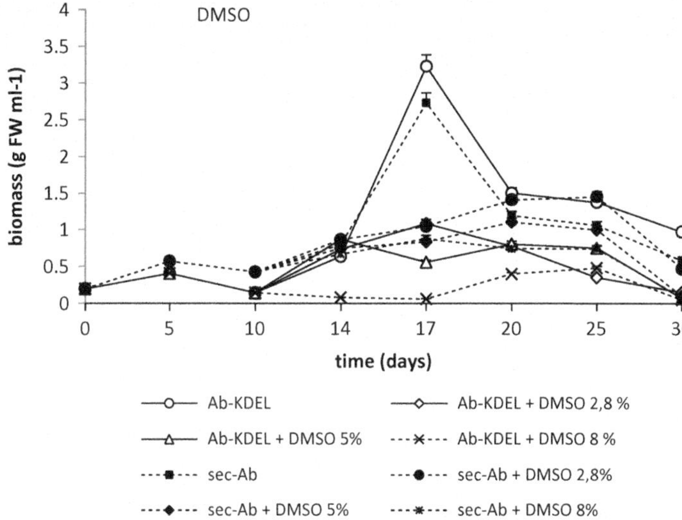

Fig. 7.5 Influence of the addition of DMSO at different concentrations (2.8 % V/V, 5 % V/V and 8 % V/V) on growth of *N. tabacum* sec-Ab and Ab-KDEL lines

7.3.2.2 Influence of Gelatine

Gelatine was added to the culture medium at three concentrations (1.0 g l^{-1}, 5.0 g l^{-1}, 9.0 g l^{-1}) at the 22nd day of culture corresponding to the highest 14D9 accumulation in non-treated hairy roots. At a concentration of 5.0 g l^{-1} gelatine had a positive effect on 14D9 concentration in the culture medium. However, it had a negative effect on growth at concentrations higher than 1.0 g l^{-1}. The maximal concentration of 14D9 in the culture medium corresponded to a concentration of gelatine of 5.0 g l^{-1}, with a 14D9 concentration of 15 μg 14D9 ml^{-1} at the 22nd day of culture. As for the %TSP in the biomass it does not show significant differences between the treatments and the control (Fig. 7.7). The positive effect of 5.0 g l^{-1} gelatine on the amount of the antibody concentration in the culture medium could be attributed to a blocking of specific interactions between the antibody and the Erlenmeyer glass surface, a competition with the substrate for the enzymes that degrades the immunoglobulin, modification of interactions among the recombinant IgG and components of the MSRT culture medium (Martínez et al. 2005, Doran 2006). The obtained results agree with those reported by Wongsamuth and Doran (1997) which demonstrated a stabilizing effect of gelatine, at concentrations of 1.0–9.0 g l^{-1}, on the heavy chain of the monoclonal antibody Guy's 13 in *N. tabacum* suspension cell cultures.

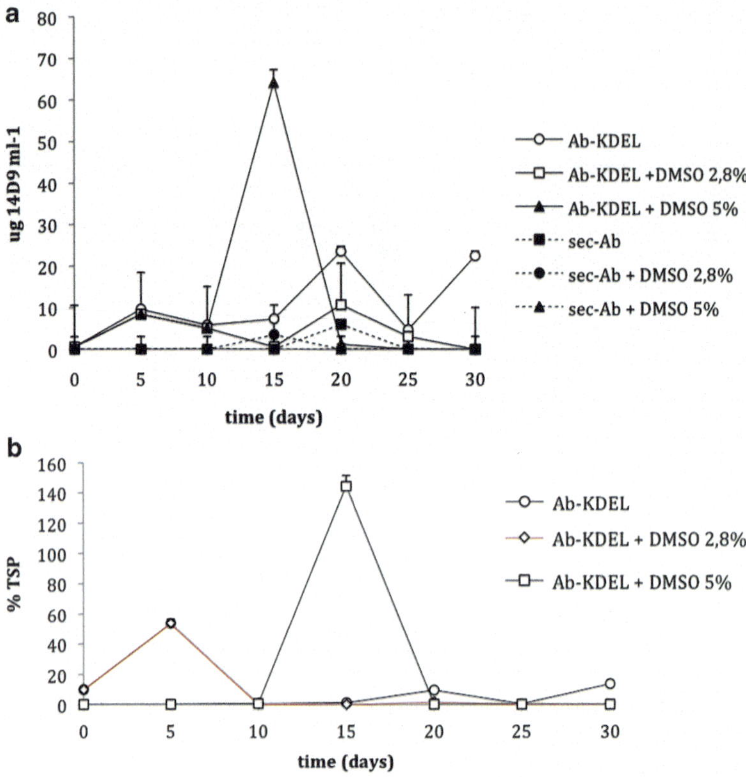

Fig. 7.6 The 14D9 content in the biomass of sec-Ab and Ab-KDEL *N. tabacum* hairy root lines, and (**b**) the total soluble protein content (%TSP) in the biomass of the Ab-KDEL *N. tabacum* hairy root line

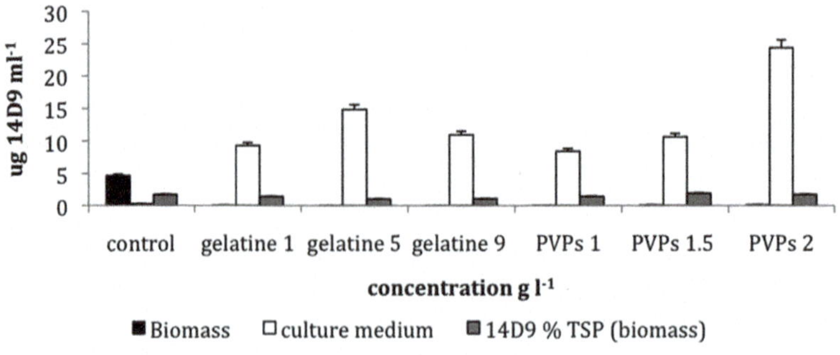

Fig. 7.7 Influence of gelatine (1 g l^{-1}, 5 g l^{-1}, 9 g l^{-1}) and PVP (1 g l^{-1}, 1.5 g l^{-1}, 2 g l^{-1}) on 14D9 amount in the biomass, the culture medium and on the 14D9 % TSP in the biomass

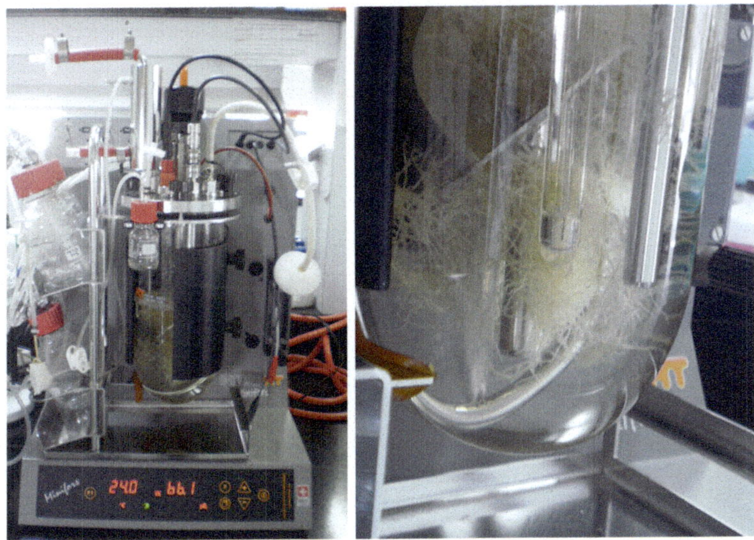

Fig. 7.8 *N. tabacum* hairy roots, line sec-Ab growing in a 2 l- bioreactor (Minifors) in MSRT media without plant growth regulators

7.3.2.3 Influence of Polyvinylpyrrolidone

Polyvinylpyrrolidone (PVP) is a versatile polymer, water soluble, with excellent colloidal, stabilizing and complexing properties, being at the same time metabolic and physiological inert (Magnuson et al. 1996). For analyzing its effect on 14D9 production, PVP was added to the Ab-KDEL hairy root cultures at three different concentrations (1 g l^{-1}, 1.5 g l^{-1} and 2.0 g l^{-1}). The effect was analyzed at the 22nd day of culture that corresponds to the highest 14D9 amount in the hairy roots without treatment. The addition of PVP at a concentration of 1.0 gl^{-1} had significant positive effect ($P<0.01$) on growth in the Ab-KDEL line (Fig. 7.7). Other researchers have reported a negative effect on growth of PVP concentrations higher than 3.0 g l^{-1} (Magnuson et al. 1996). On the other hand, 14D9 concentration increased significantly ($P<0.01$) in the culture medium (24 μg ml^{-1}) at a PVP concentration of 2.0 g l^{-1} (Martínez et al. 2005). Maybe the protective effect of PVP avoids the antibody denaturalization or its adherence to the contact surfaces (Doran 2006) (Fig. 7.8).

7.4 Production of 14D9 in Cell Suspension Cultures

From a biotechnological point of view it is desirable to perform a productive process in suspended cells for the more straight control of the environmental conditions at cellular level. We have established cell suspension cultures of both 14D9 expressing-recombinant lines (sec-Ab and Ab-KDEL).

Fig. 7.9 *N. tabacum* calli established in MSRT medium with NAA: kin (2:0.2 mg ml^{-1}) as plant growth regulators

7.4.1 Establishment of Calli Cultures

Pieces of leafs of transgenic plants expressing the antibody 14D9 in both versions (sec-Ab and Ab-KDEL) were transferred to Petri dishes containing MSRT solid medium (8 g l^{-1} agar, 30 g l^{-1} sucrose) with the addition of one of NAA: kinetin (2 mg l^{-1}: kinetin 0.2 mg l^{-1}). The cultures were kept in a chamber at 24 ± 2 °C, and a 16-hs photoperiod given by fluorescent lamps (irradiance 1.8 w m^{-1} seg^{-1}). After a couple of weeks, calli appeared in the wounded surfaces. They were separated, transferred to fresh media, and subcultured monthly (Fig. 7.9). At each subculture, samples were taken by triplicate to measure 14D9. Using this methodology, the more productive and stable lines are maintained since 2005.

7.4.2 Establishment of Cell Suspension Cultures

Friable calli from the lines sec-Ab and Ab-KDEL were transferred to Erlenmeyer flasks (5 % W/V) with MSRT liquid medium plus sucrose (30 g l^{-1}) and NAA: kinetin (2 mg l^{-1}: 0.2 mg l^{-1}). The suspended cells (Fig. 7.10) were transferred to identical fresh media (maintenance medium) every 20 days, and maintained at 24 ± 2 °C, a 16-hs photoperiod (irradiance 1.8 w m^{-1} seg^{-1}) in an orbital shaker at 100 rpm (López et al. 2010).

Fig. 7.10 *N. tabacum* cell suspension culture, line Ab-KDEL

7.4.3 Kinetic of Growth and 14D9 Expression in N. tabacum *Cell Suspension Lines sec-Ab and Ab-KDEL*

For analyzing the profile of growth and 14D9 production of cell suspension cultures, 2.0 g FW of an 8-day-old culture were transferred to 150 ml-Erlenmeyer flasks with 40 ml of MSRT maintenance medium and cultured in the same conditions described in Sect. 7.4.2 during 30 days. Samples were taken every 5 days for determining cell growth and to quantify 14D9.

The specific growth rate and the amount of 14D9 in the biomass were significant higher in the Ab-KDEL line than in the sec-Ab and wild type lines. However, there are no significant differences in maximal biomass among the lines (Fig. 7.11, Table 7.2).

7.4.4 Influence of DMSO

DMSO was used in an attempt to increase the amount of 14D9 in the culture medium because it was reported that it can produce a stabilizing effect on proteins by forming hydrogen bonds with the protein proton donor groups (Henderson et al. 1975) acting as a chemical chaperone (Welch and Brown 1996).

To express 14D9, DMSO was added at the 3rd day of *N. tabacum* lines sec-Ab and Ab-KDEL cultures, at different concentrations (1.5, 2.0, 2.5 %). At all the concentrations tested DMSO has a negative effect ($P<0.05$) on growth. At a 1.5 % concentration DMSO produced an 8-fold increase of 14D9 in the culture medium at

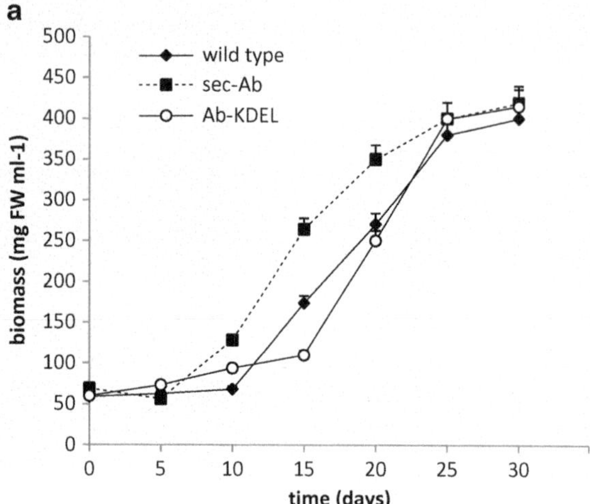

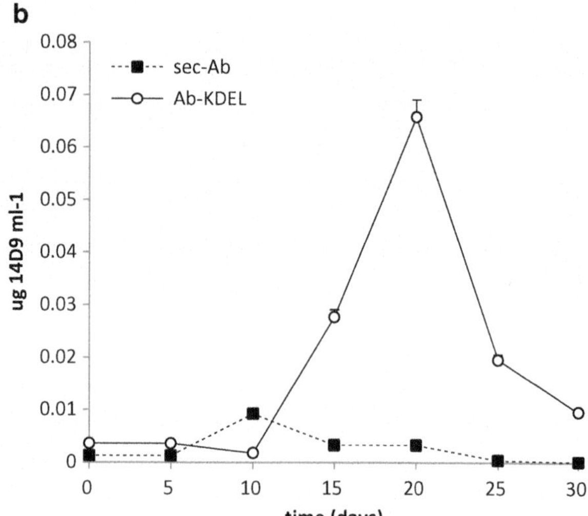

Fig. 7.11 (**a**) Kinetic of growth of *N. tabacum* wild type, sec-Ab and Ab-KDEL cell suspension cultures, (**b**) 14D9 accumulation in the biomass of both transgenic lines

Table 7.2 Parameters of growth and 14D9 production on *N. tabacum* wild type, sec-Ab and Ab-KDEL lines

N. tabacum line	μ (day^{-1})	td	Maximal biomass (mg ml^{-1} FW)	14D9 (μg 14D9 g^{-1} FW)
Wild type	0.086	8.05	400	–
sec-Ab	0.069	10.7	419	0.012
Ab-KDEL	0.149	4.62	415	0.470

the 10th day of culture in the line sec-Ab, while at a 2.0 % concentration there is a 5-fold increase in the culture medium at the 11st day of culture in the line Ab-KDEL. Total protein and 14D9 content in the biomass did not increase.

7.4.5 *Effect of Gelatine and PVP on Growth and 14D9 Yield*

Protein stabilizing agents and protease inhibitors are employed to increase the amount of recombinant proteins in cultures. Some of the protein stabilizers used are gelatine, bovine seroalbumin (BSA) (for protein-based stabilization), mannitol (to regulate the osmotic pressure of the medium to minimize cell lysis), PEG and other polymers (as stabilizing agents and to protect proteins from plant cell denaturing agents) (Benchabane et al. 2008; Guillon et al. 2006; Magnuson et al. 1996; Sharp and Doran 2001).

Different concentrations of gelatine (1.0, 5.0 or 9.0 g l^{-1}) or PVP (1.0, 1.5 or 2.0 g l^{-1}) were added to the culture medium at the first day of culture. Samples were taken by triplicate at the 22nd day of culture to measure FW, DW, and antibody accumulation in the biomass and in the culture media. The culture conditions were also the same as described above. When PVP was added at 1.0 g l^{-1} and 1.5 g l^{-1} concentration, the antibody in the biomass increased from 1.4 % TSP to 4.8 % TSP and 2.99 % TSP respectively in Ab-KDEL line. In the case of the sec-Ab line the 14D9 yield increased from 0.17 % TSP to 1.16 % TSP with a PVP concentration of 1.5 g l^{-1} (Fig. 7.12).

As for the amount of antibody in the culture medium, an increase from 0.06 %TSP (control) to 0.18 %TSP and 0.15 % TSP was obtained with 1.0 and 1.5 g l^{-1}

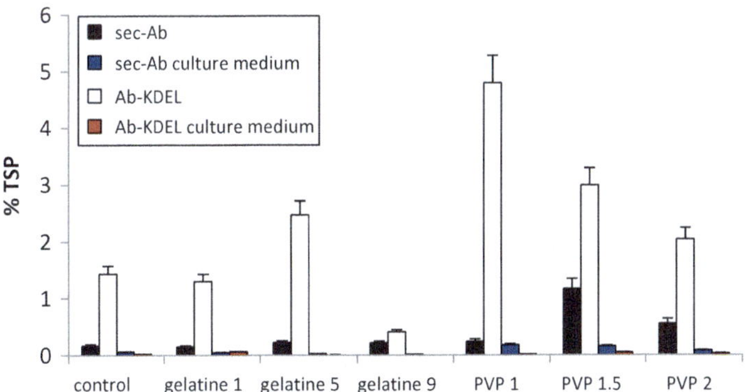

Fig. 7.12 Effect of gelatine (1, 5 and 9 g l^{-1}) and PVP (1, 1.5 and 2 g l^{-1}) on 14D9 accumulation (as % of total soluble protein) in the biomass and in the culture medium of *N. tabacum* sec-Ab and Ab-KDEL transgenic lines

PVP respectively for sec-Ab cell suspension cultures; no significant changes were detected in the culture media of the Ab-KDEL cell suspension line. Concerning the influence of gelatine, at a concentration of 5.0 g l^{-1} the antibody yield increased from 1.43 % TSP to 2.47 % TSP in Ab-KDEL cell suspensions. No significant influence of levels in the biomass of the sec-Ab line, or in the culture medium was determined (López et al. 2010).

The beneficial effect of PVP is attributed to its excellent complexing, stabilizing and colloidal properties, while it is metabolically and physiologically inert, or for being taken as sacrificial substrate for protease activity. Moreover, PVP would be the preferred additive in a production process for being a non-human and non-animal source.

Protease concentration in the biomass was 1.0 U g FW^{-1} h^{-1}, and remains constant along the culture, whereas in the media the protease activity increased after the 10th day in culture to a maximal level of 0.16 U g FW^{-1} h^{-1} until the end of the culture.

Our results are partially in agreement with those reported for other authors (Sharp and Doran 2001; Wongsamuth and Doran 1997) when expressing the secreted version of the monoclonal antibody Guy's13 in *N. tabacum* cell suspensions and hairy roots. They did not found a significant influence of PVP on antibody accumulation in the biomass of cell suspensions and have established that PVP and gelatine diminishes antibody content in hairy root biomass. However, they have reported an increase of total antibody accumulation both in cell suspension as in hairy roots when PVP or gelatine were added at a concentration of 1.5 g l^{-1} or 9.0 g l^{-1} respectively due to an increase in antibody accumulation in the culture medium (López et al. 2007).

7.4.6 *Kinetic of Growth and 14D9 Expression After Three Years in Culture of the sec-Ab and Ab-KDEL Lines*

After three years in culture the profile of growth and 14D9 production in the sec-Ab and Ab-KDEL lines was studied and compared to those of the first year in culture. As is shown in Fig. 7.13, a reduction of the lag phase from 10 to 3 days is evident. Also the specific growth rate (μ), and the amount of 14D9 in the biomass increased (Table 7.3). No significant differences ($P>0.05$) were found in the final biomass among the transgenic and the wild type lines demonstrating that the expression of the heterologous protein do not has a negative consequence on cell growth (López et al. 2010).

The amount of sucrose in the culture medium was 11.4 g l^{-1} in the wild type at the end of the culture, while for the sec-Ab and Ab-KDEL lines, sucrose was totally consumed at the 10th day and at the 28th day of culture respectively. The maximal biomass yield related to the consumed substrate (Yx/s) was in all cases into the average values for plant cells (0.3–0.7 g DW g $sugar^{-1}$) (Payne et al. 1991).

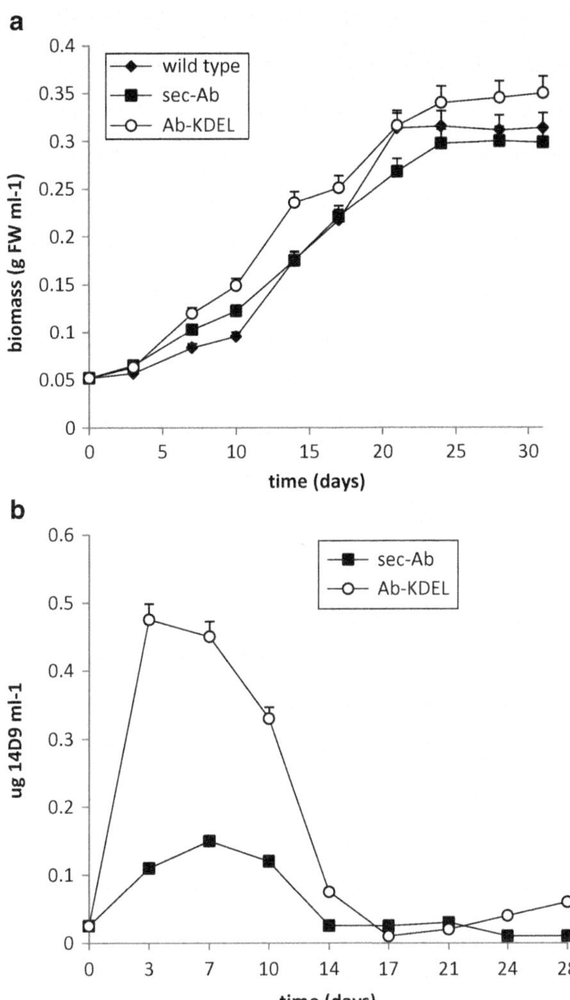

Fig. 7.13 (**a**) Kinetic of growth of *N. tabacum* wild type, sec-Ab and Ab-KDEL lines and (**b**) 14D9 amount in the biomass of the transgenic lines after 3 years in culture

Table 7.3 Parameters of growth and 14D9 production in *N. tabacum* wild type, sec-Ab and Ab-KDEL lines after 3 years in culture

Line	μ (day^{-1})	td (days)	GI	Final biomass (g FW)	Yx/s (g DW g consumed sucrose^{-1})	μg 14D9 g FW^{-1}
Wild type	0.098	7.07	6.65	13.73 ± 1.30	0.711 ± 0.123	–
Sec-Ab	0.086	8.06	5.88	12.20 ± 1.20	0.419 ± 0.011	0.73 ± 0.05
Ab-KDEL	0.123	5.64	7.13	14.79 ± 2.53	0.467 ± 0.049	8.21 ± 0.82

There is a threefold increase of 14D9 level in the Ab-KDEL line respect to the sec-Ab line, and a twofold increase respect to leaves of the whole transformed plant (Petruccelli et al. 2006). The differences in 14D9 levels between the sec-Ab and Ab-KDEL line confirmed the protector effect of the ER-retention on antibody accumulation. The more oxidant environment, with less proteases and high amount of chaperones like BiP, promotes the correct folding and decrease the antibody degradation (Schouten et al. 1997). The fall in 14D9 amount at the end of the cultures in both transgenic lines can be attributed to a diminution in the transgene expression (RNAm), by an increase in the activity of proteases not detected by the test of azo-caseine digestion, and by the formation of 14D9 aggregates that can produce false negatives (Sharp and Doran 2001; Schouten et al. 1996).

7.5 Influence of Mannitol on Growth and 14D9 Expression

Mannitol produces an osmotic stress in plant cells that alter their gene expression and increase the proteins synthesis (Soderquist and Lee 2005). Also, it was described a certain stabilizing effect on proteins (Wimmer et al. 1997). We analyzed the effect of mannitol (90 g l^{-1}) on cell growth and 14D9 yield in *N. tabacum* suspended cells (Fig. 7.8). Fresh weight and μ decreased significantly ($P<0.05$) both in the sec-Ab and Ab-KDEL, an effect derived from the high osmolality of the culture medium (López et al. 2010). That negative effect of the osmotic stress on cell growth was also reported for hybridomes and animal cultures, and in *N. tabacum* suspended cells expressing an IgG in the presence of 17 g l^{-1} mannitol (Tsoi and Doran 2002).

The utilization of sucrose was also affected. The three lines (sec-Ab, Ab-KDEL, wild type) consumed sucrose during growth. Sucrose is hydrolyzed to glucose and fructose by the activity of invertases associated to the plant cell walls; those sugars are used as carbon source by plant cells. Glucose accumulated at the beginning of the culture when sucrose was abundant and the hydrolysis of sucrose and glucose production was higher than the glucose consumed by plant cells. When sucrose decreased in the culture medium, the glucose concentration remained constant because the rate of production and consumption are balanced, finally glucose concentration decreased when the sucrose in the culture medium is almost zero. In the presence of an osmolite as mannitol the consumption of sucrose is lower; perhaps mannitol can be a supplementary carbon source, as happen in suspension cell cultures of celery that use mannitol as carbon source (Table 7.4).

Table 7.4 Influence of mannitol (90 g l^{-1}) on specific growth rate (μ), growth (final biomass) and antibody content

N. tabacum line	μ (day^{-1})	Final biomass (g l^{-1})	14D9 (μg ml^{-1})
Sec-Ab	0.086 ± 0.002	12.20 ± 1.20	0.018 ± 0.006
Sec-Ab + mannitol	0.052 ± 0.006	9.42 ± 0.79	1.96 ± 0.20
Ab-KDEL	0.123 ± 0.0053	14.79 ± 2.53	0.20 ± 0.082
Ab-KDEL + mannitol	0.11 ± 0.030	13.00 ± 1.25	2.31 ± 0.18

Total protein content and 14D9 increased in the biomass ($P<0.05$) but not in the culture medium ($P>0.05$). In the biomass the values were twofold higher respect to the control without significant differences between the transgenic lines.

The level of antibody in the culture media was not affected by the presence of mannitol, discarding the stabilizing effect of mannitol on proteins. The antibody secreted to the culture medium was 0.20 % of the amount in the biomass for the sec-Ab line and 0.02 % of the amount in the biomass for the Ab-KDEL line. In the sec-Ab line, 14D9 would be retained in the apoplast, the space between the cell membrane cell wall. The exclusion size of the cell wall pores is approximately 20–30 kDa (Carpita et al. 1979) although, in some cases proteins of 50 kDa, and even whole antibodies of 150 kDa, can pass through the cell wall (Hein et al. 1991). In our experiments 14D9 is apparently retained in the apoplast (sec-Ab) or in the ER (Ab-KDEL). The low levels of 14D9 in the culture medium can be a consequence of an inefficient retention in the ER that permit an antibody escape to the apoplast (Petruccelli et al. 2006), or to cell lysis and the consequent liberation of the antibody to the culture medium. The loss of 14D9 in the culture medium at the end of the culture can be attributed to its adsorption to the vessel walls (Doran 2006), to the degradation by proteases, which were detected at lower activities than in the biomass, or to a combination of both phenomena. The direct ELISA demonstrated the ability of the recombinant 14D9 of recognizing and binding its haptene.

7.6 Large-Scale Production in a 2-l Bioreactor

Suspended cells of the Ab-KDEL line (inoculum size 1 % or 5 %, by W/V) were transferred to a 2-l stirred-tank bioreactor (Minifors, Infors HT®, Switzerland) containing 1.0 l of MSRT medium with NAA: kinetin (20:02 mg l^{-1}) as PGR (Fig. 7.14). A marine propeller produced the mechanical agitation and a porous metal sparger provided the bubble aeration. Culture conditions were 100 rpm, 0.1 vvm, and 24 ± 2 °C. Relative partial O2 pressure and pH were monitored on-line. Samples were taken every 2 days during a culture period of 15 days, for measuring FW, DW, and 14D9 concentration. Three replicates were used in all analytical determinations. Analysis of variance (ANOVA) was performed for each test. Table 7.5 shows the parameters of growth and antibody content of *N. tabacum* in the 2-l bioreactor.

Regarding the GI and the antibody content, both were higher with a 1 % P/V than with a 5 % P/V inoculum size. As for biomass yield respect to the carbon source (YX/S), it was significant lower in the bioreactor when the inoculum size was 5 % V/V respect to the inoculum size of 1 % V/V probably due to a nutrient limitation derived from the highest cell density in the culture (Marconi and Alvarez 2014b).

Comparing the process performed in Erlenmeyer flasks and the bioreactor, the antibody yield was significantly higher in the 2-l bioreactor than in Erlenmeyer flasks. Those results are predictable considering the highest availability of nutrients and the earlier antibody peak production attained in the bioreactor (between the

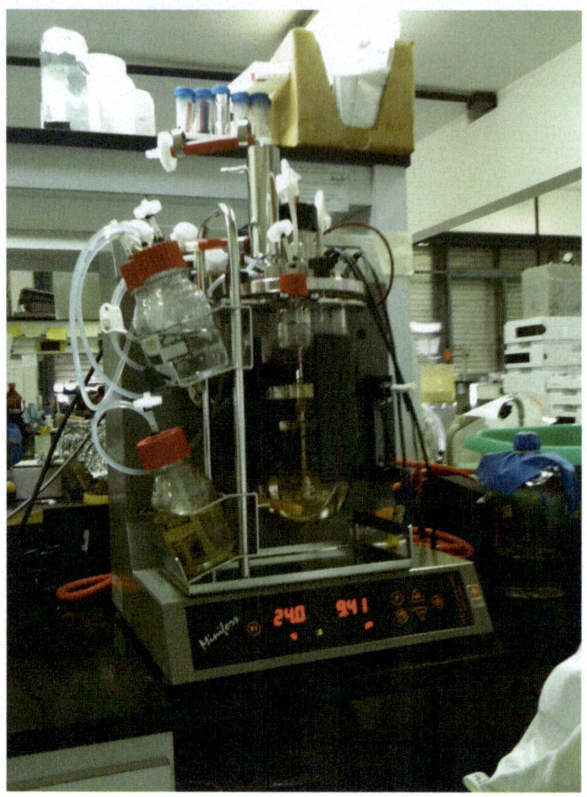

Fig. 7.14 2-l bioreactor inoculated with MSRT culture media without plant growth regulators for growing *N. tabacum* cell suspension culture line Ab-KDEL

Table 7.5 Parameters of growth and 14D9 production of *N. tabacum* Ab-KDEL line in a 2-l Minifors bioreactor at two inoculum sizes (1 % P/V and 5 % P/V)

Inoculum size (% P/V)	μ (day^{-1})	Final biomass (mg FW ml^{-1})	14D9 (μg g FW^{-1})
1	0.084	70.9	0.66
5	0.291	177.91	0.21

days 5 to 10 of culture) than in Erlenmeyer flasks (between the days 10 to 20 of culture). Comparing the performance of the Ab-KDEL cell suspension culture in the bioreactor with that of hairy roots in Erlenmeyer flaks, yield expressed as % TSP was also higher in suspended cells growing in the bioreactor. However, the productivity values were considerable lower than those of hairy roots that would be attributable to the fast growth and high productive biomass characteristics of hairy root cultures.

7.7 Conclusive Remarks

The 14D9 catalytic antibody was expressed in *N. tabacum* hairy roots and undifferentiated *in vitro* cultures. Figure 7.15 summarizes the procedures followed to optimize yields. Both systems (hairy roots and cell suspension cultures) were a feasible antibody productive platform. The presence of the KDEL ER-retention signal and the use of PVP at a 1 % g l^{-1} concentration resulted in the highest antibody yields in Erlenmeyer flasks. The scale up of Ab-KDEL suspended cells to a 2-l stirred-tank bioreactor increased 14D9 yield (expressed as % TSP) but productivity remained higher in hairy root cultures due to their highest productive biomass.

Comparing 14D9 yields obtained in all the platforms and systems tested, we can conclude that hairy roots growing in Erlenmeyer flasks and Ab-KDEL cell suspension cultures growing in a 2-l bioreactor gave the highest 14D9 yields. Further work is being conducted in order to optimize 14D9 production by *N. tabacum* cell suspension and hairy root cultures in a bioreactor.

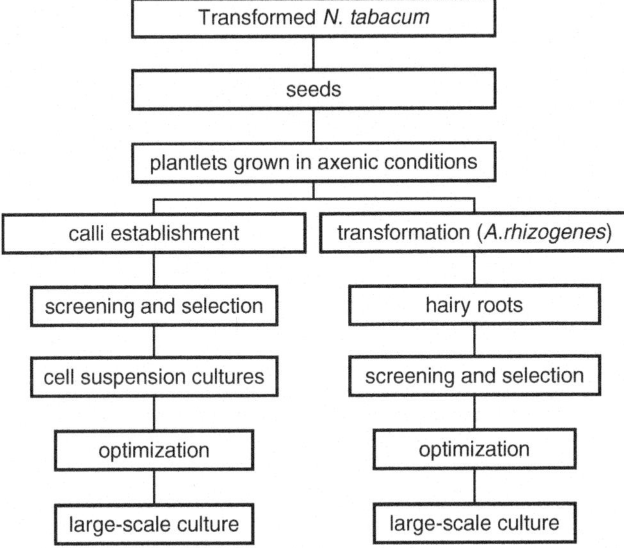

Fig. 7.15 General procedure followed to establish *N. tabacum in vitro* cultures for producing 14D9

References

Alvarez MA, Rodríguez Talou J, Paniego N, Giulietti AM (1994) Solasodine production in transformed cultures (roots and shoots) of *Solanum eleagnifolium* Cav. Biotechnol Lett 16(4): 393–396

Benchabane M, Goulet C, Rivard D, Faye L, Gomord V, Michaud D (2008) Preventing unintended proteolysis in plant protein biofactories. Plant Biotechnol J 6:633–648

Brown CR, Hong-Brown LQ, Biwersi J, Wekman AS, Welch WJ (1996) Chemical chaperones correct the mutant phenotype of the delta F508 cystic fibrosis transmembrane conductance regulator protein. Cell Stress Chaperones 1(2):117–125

Carpita N, Sabularse D, Montezinos D, Delmer DP (1979) Determination of the pore size of cell walls of living plant cells. Science 205:1144–1147

Desai PN, Shrivastava N, Padh H (2010) Production of heterologous proteins in plants: strategies for optimal expression. Biotechnol Adv 28:427–435

Doran PM (2006) Loss of secreted antibody from transgenic plant tissue culture due to surface adsorption. J Biotechnol 122:39–54

Egelkrout E, Rajan V, Howard JA (2012) Overproduction of recombinant proteins in plants. Plant Sci 184:83–101

Guillon S, Guiller JT, Pati PK, Rideau M, Gantet P (2006) Hairy root research: recent scenario and exciting prospects. Curr Opin Plant Biol 9:341–346

Hasserodt J (1999) Organic synthesis supported by antibody catalysis. New Tools Synth 12: 2007–2022

Hein MB, Tang Y, McLeod DA, Janda KD, Hiatt A (1991) Evaluation of immunoglobulins from plant cells. Biotechnology Progress 7:455–461

Henderson TR, Henderson RF, York JL (1975) Effects of dimethyl sulfoxide on subunit proteins. Ann N Y Acad Sci 243:38–53

Hiatt A, Cafferkey R, Bodwish K (1989) Production of antibodies in plants: status after twenty years. Plant Biotechnol J 8:529–563

Horsch R, Fraley R, Rogers S, Sanders P, Lloyd A, Hoffman N (1984) Inheritance of functional foreign genes in plants. Science 223:496–498

Jahangiri GK, Reymond JL (1994) Antibody-catalyzed hydrolysis of enol ethers. 2. Structure of the antibody-transition state complex and origin of the enantioselectivity. J Am Chem Soc 116: 11264–11274

López J, Parma Y, Meroño T, Petruccelli S, Marconi P, Pitta-Alvarez S, Alvarez MA (2007) Producción del anticuerpo recombinante 14D9 en suspensiones celulares de *Nicotiana tabacum*. Boletín Latinoam Caribe Plantas Medicinales Aromáticas (BLACPMA) 6(6):399–400

López J, Lencina F, Petruccelli S, Marconi P, Alvarez MA (2010) Influence of the KDEL signal, DMSO and mannitol on the production of the recombinant antibody 14D9 by long-term *Nicotiana tabacum* cell suspension culture. Plant Cell Tissue Organ Cult 103:307–314

Magnuson NS, Linzmaier PM, Gao JW, Reeves R, Gynheung AN, Lee JM (1996) Enhanced recovery of a secreted mammalian protein from suspension culture of genetically modified tobacco cells. Protein Expr Purif 7(2):220–228

Marconi PL, Alvarez MA (2014a) State of the art on plant-made single-domain antibodies. J Immunol Techn Infect Dis, in press. doi: 10.4172/2329-9541.1000125

Marconi PL, Alvarez MA (2014b) The expression of the 14D9 antibody in suspended cells of *Nicotiana tabacum* cultures increased by the addition of protein stabilizers and by transference from Erlenmeyer flasks to a 2-l bioreactor. Biotechnol Prog, in press. doi: 10.1002/btpr.1940

Martínez CA, Petruccelli S, Giulietti AM, Alvarez MA (2005) Expression of the antibody 14D9 in *Nicotiana tabacum* hairy roots. Electron J Biotechnol 8(2):170–176

Paul M, Ma JKC (2011) Plant-made pharmaceuticals: leading products and production platforms. Biotechnol Appl Biochem 58:58–67

Payne GF, Bringi V, Prince C, Shuler ML (1991) The quest for commercial production of chemicals from plant cell culture. In: Payne GF, Bringi V, Prince C, Shuler ML (eds) Plant cell and tissue culture in liquid systems. Hanser Publishers, Munich, pp 1–10

Petruccelli S, Otegui MS, Lareu F, Trandinhthanhlien O, Fichette AC et al (2006) A KDEL-tagged monoclonal antibody is efficiently retained in the endoplasmic reticulum in leaves, but is both partially secreted and sorted to protein storage vacuoles in seeds. Plant Biotechnol J 4:511–527

Reymond JL (1999) Catalytic antibodies for organic synthesis. In: Biocatalysis-from discovery to application, Topics in current chemistry. Springer, Berlin/Heidelberg, pp 59–93

Reymond JL, Janda KD, Lerner RA (1991) Antibody catalysis of glycosidic bond hydrolysis. Angew Chem Int Ed 30:1711–1714

Reymond JL, Jahangiri GK, Stoudt C, Lerner RA (1993) Antibody catalyzed hydrolysis of enol ethers. J Am Chem Soc 115(10):3909–3917

Reymond JL, Reber RA, Lerner RA (1994) Enantioselective, multi-Gram scale synthesis with a catalytic antibody. Angew Chem Int Ed Engl 33:475–477

Schouten A, Roosien J, de Boer JM, Wilmink A, Rosso MN et al. (1997) Improving scFv antibody expression levels in the plant cytosol. FEBS Letters 415:235–241

Schouten A, Roosien J, van Engelen FA, de Jong GA, Borst-Vrenssen AWM, Zilverentant JF (1996) The C-terminal KDEL sequence increases the expression level of a single chain antibody designed to be targeted to both the cytosol and the secretory pathway in transgenic tobacco. Plant Mol Biol 30:781–793

Sharp JM, Doran PM (2001) Strategies for enhancing monoclonal antibody accumulation in plant cell and organ cultures. Biotechnol Prog 17(6):979–992

Sinha SC, Keinan E, Reymond HL (1993) Antibody-catalyzed reversal of chemoselectivity. Proc Natl Acad Sci USA 90:11910–11913

Soderquist RG, Lee JM (2005) Enhanced production of recombinant proteins from plant cells by the application of osmotic stress and protein stabilization. Plant Cell Rep 24:127–132

Tanaka F (2002) Catalytic antibodies as designer proteases and esterases. Chem Rev 102: 4885–4906

Tsoi BMY, Doran P (2002) Effect of medium properties and additives on antibody stability and accumulation in suspended plant cell cultures. Biotechnol Appl Biochem 35:171–180

Ulrich HD, Patten PA, Yang PL, Romesberg FE, Schultz PG (1995) Expression studies of catalytic antibodies. Proc Natl Acad Sci USA 92:11907–11911

Welch WJ, Brown CR (1996) Influence of molecular and chemical chaperones on protein folding. Cell Stress Chaperones 1:109–115

Whal MF, An GHA, Lee JM (1995) Effects of dimethyl sulfoxide on heavy chain monoclonal antibody production from plant cell culture. Biotechnol Lett 17(5):463–468

Wimmer R, Olsson M, Petersen MTN, Hattikaul R, Petersen SB, Muller N (1997) Towards a molecular level understanding of protein stabilization: the interaction between lysozyme and sorbitol. J Biotechnol 55:85–100

Wongsamuth R, Doran P (1997) Production of monoclonal antibodies by tobacco hairy roots. Biotechnol Bioeng 54(5):401–415

Yoshida K, Matsui T, Shinmyo S (2004) The plant vesicular transport engineering for production of useful recombinant proteins. J Mol Catal B: Enzym 28:167–171

Zheng L, Baumann U, Reymond JL (2003) Production of a functional catalytic antibody ScFv-NusA fusion protein in bacterial cytoplasm. J Biochem 133:577–581

Zheng L, Baumann U, Reymond JL (2004) Molecular mechanism of enantioselective proton transfer to carbon in catalytic antibody 14D9. Proc Natl Acad Sci USA 101(10):3387–3392

Chapter 8
Expression of the Potentially Immunogenic Truncated Glycoprotein E2 (From Viral Bovine Diarrhoea Virus) in *Nicotiana tabacum*

Abstract Viral Bovine Diarrhoea (VBD) is a contagious infection disease caused by the viral bovine diarrhoea virus (BVDV) producing high economic losses for its effects in the respiratory and reproductive system. Besides bovines, BVDV also affects sheep, goats, and wild ruminants, which are potential reservoirs of the virus.

The main purpose of a program for controlling BVDV should be the prevention of the foetal infections, avoiding the birth of persistently infected animals (PIA), and preventing acute infections. Those goals can be achieved preventing the exposure to the virus by removing the PIA from the herd, and by increasing the immunity to BVDV through vaccination.

The glycoprotein E2, as the principal immunogen of the BVDV, is the main target for neutralizing antibodies. In the last decade, studies were performed to develop vaccines with a recombinant E2 of prokaryotic and eukaryotic origin. E2 was expressed in *E. coli*, mammal's cells through the vaccinia system, insect cells and plants. In insect cells and plants the E2 expressed was able to induce specific neutralizing antibodies being potential production platforms for the development of a new generation of vaccines.

Keywords Viral bovine diarrhoea • Glycoprotein E2-molecular farming • *Nicotiana tabacum*-plant transformation • Plant-made proteins

8.1 Introduction

Vaccines are fundamental tool to fight the majority of infectious diseases; therefore, the recent outbreaks have triggered huge efforts to R&D of new vaccines. The limiting steps are the considerable extent of production processes, high costs, and the need of a fast response (Alvarez 2012). Besides the traditional production methods, protein immunogens to produce subunit vaccines are being produced in yeasts, insect cells, plant cell cultures and transgenic plants with different degree of success (Bolin and Ridpath 1996; Smith et al. 2011; Xu et al. 2012; Hefferon 2013).

The Bovine Viral Disease produces considerable economical losses when outbreaks in cattle, which put in focus the research for new generation vaccines.

M.A. Alvarez, *Plant Biotechnology for Health: From Secondary Metabolites to Molecular Farming*, DOI 10.1007/978-3-319-05771-2_8

8.2 Viral Bovine Diarrhoea

Viral Bovine Diarrhoea (VBD) is a contagious infection disease caused by the viral bovine diarrhoea virus (BVDV), a virus of the genus Pestivirus (family Flaviviridae). BVDV affects bovines, causing high economic losses for its activity in the respiratory and reproductive system (Lértora 2003). Besides, BVDV affects sheep, goats and wild ruminants, which are potential reservoirs of the virus (Bracamonte et al. 2006). The virus was described for the first time in 1946 in the United States of America, soon after it was detected in many countries as England, Australia, Kenya, Germany and Argentina.

8.3 The Viral Bovine Diarrhoea Virus (BVDV)

The BVDV is a positive chain RNA virus, antigenically related to other Pestivirus as the Classic Swine Fever Virus (CSFV) and the ovine Border Disease Virus (BDV) (Pringle 1999). Besides the antigenic variation, the isolates of BVDV can be classified in cytopathic (cp) and non-cytopathic (ncp) strains according to their ability to induce cytopathic effects in cell cultures. Several BVDV strains maintain their cytopathogenicity in a broad spectrum of host tissues, while others change during the passages through different host cell systems.

8.4 Vaccines

The main purpose of a program for controlling BVDV should be the prevention of the foetal infections, avoiding the birth of persistently infected animals (PIA), and preventing acute infections (Harkness 1987). Those goals can be achieved preventing the exposure to the virus by removing the PIA from the herd, and by increasing the immunity to BVDV through vaccination.

The vaccines currently in the market do not produce a significant reduction of BVDV prevalence, which drives research of more effective and secure vaccines. There are two types of commercial vaccines: modified-live vaccines (MLV) and inactivated vaccines (IV) (Bolin and Ridpath 1995). In general, in both cases, the neutralizing titres are in the order of 1:2,000 (Fulton and Burge 2001).

Although it was demonstrated that passively acquired antibodies could protect against the challenge with BVDV (Bolin and Ridpath 1995), the relationship between the neutralizing titre and the protective effect it is not yet clear (van Oirschot et al. 1999). However, the study of the efficacy of vaccines after vaccination lies in the sero-conversion test and other measurable parameters, as rectal temperature and leucopoenia, after a viral challenge in vaccinated animals.

BVDV also suppresses the immune system facilitating other pathogens entry. Consequently, a multi-vaccine complex with other antigens as herpes bovine-1 virus, parainfluenza-3 virus, respiratory syncytial virus, and different species of *Pasteurella* and *Haemophilius*, is commonly intramuscular or subcutaneously administered (van Oirschot et al. 1999).

8.5 Role of the Adjuvant

Adjuvants, which increase the response to antigens, are vaccine components that contribute to the delivery of the immunogen and to the maintenance and modulation of the immune response (Aucouturier et al. 2001). Besides, as they improve the vaccine efficacy, they reduce the amount of antigen required in the formulation. Live-viral traditional vaccines do not require the addition of an adjuvant. The modern recombinant vaccines, with highly purified synthetic antigens, need adjuvants for inducing the long-term protective immune response. Aluminium salts are the most employed adjuvants in human vaccines, but they are weak and promote mainly the induction of neutralizing antibodies than cellular immunity. Other adjuvants in use are oily emulsions, bacterial products (B subunit of the *Vibrio colerae* toxin), viral products (virus-like particles), plant secondary metabolites (saponins), biodegradable particles (liposome), etc. (Lambrecht et al. 2009; Reed et al. 2008; Aucouturier et al. 2001). Adjuvants may produce their effects through different mechanisms, recruiting innate immunity as a consequence of its deposit in the tissues, which results in a slow release of antigens from the injection site. Or, by maintenance of the antigen appropriate conformation through specific targeting mechanisms. Adjuvants should be chosen to optimise the required response with minimal side effects (WHO guidelines 2013; O'Hagan and Derek 2007).

8.6 The Glycoprotein E2

8.6.1 Antigenic Structure of the Glycoprotein E2

The BVDV glycoprotein E2 is the most important inductor of neutralizing antibodies (Donis et al. 1988; Magar et al. 1988; Weiland et al. 1999; Wensvoort and Terpstra 1988). The precise localization of its antigenic determinants and the relationship among them are not completely known. Some researchers have described possible epitope localization based on the reactivity patterns with monoclonal antibodies (Toth et al. 1999; Paton et al. 1992; Xue et al. 1990; Moennig et al. 1989; Corapi et al. 1988; Bolin et al. 1988). Not all the antigenic sites contribute to the neutralization of antibodies, but several BVDV-1 strains share an immunodominant neutralizing antigenic domain to which most of the neutralizing monoclonal antibodies bind (Paton et al. 1992). That domain is composed of several epitopes that map to the N-terminal region, whose conformation is probably defined by discontinuous aminoacid sequences. Those epitopes were also described for CSFV

(Yu et al. 1994; van Rijn et al. 1993). Two other epitopes were localized, one that is a fragment containing the last 90 aminoacids of the C- terminal end and a discontinuous epitope whose formation would depend on the expression of the whole protein (Toth et al. 1999). At least another lineal epitope was reported (Deregt et al. 1998a; Yu et al. 1996).

Differently from BVDV-1 where the conserved neutralizing epitopes map in a dominant immunodomains, in BVDV-2 the neutralizing antibodies bind to highly conserved epitopes in three antigenic domains (Deregt et al. 1998b). On the other hand, Ciulli et al. (2009) have reported the presence of an antigenic site in a population of viral variants with a low degree of antigenicity that could avoid the host immune system.

E2 forms homodimers and heterodimers by means of disulfide bonds with E1 (Branza-Nichita et al. 2001; Durantel et al. 2001; Thiel et al. 1991; Weiland et al. 1999) and remains associated to the Golgi membranes (Greiser-Wilke et al. 1991; Grummer et al. 2001; Weiland et al. 1999). Analysis of the crystal structure of E2 has revealed unique protein architecture with three domains (I, II and III), two Ig-like domains followed by an extended β- stranded domain with other fold. An essential role of His762 in pH sensing was proposed. Two Antigenic domains were mapped in domain I, two in domain II, and none in domain III suggesting that the latter is not exposed on the viral surface. One face of E2 appears to be solvent exposed while the opposite one is probable in the interface with E1 (Li et al. 2013). On the other hand, the absence of a class II fusion protein fold in E2, and the disorder of the N-terminal domain at low pH were reported (El Omari et al. 2013).

8.6.2 The BVDV Recombinant E2 Glycoprotein as Immunogen in Alternative Production Platforms

The glycoprotein E2, as the principal immunogen of the BVDV, is the main target for neutralizing antibodies. In the last decade, studies were performed to develop vaccines with a recombinant E2 of prokaryotic and eukaryotic origin.

The E2 expressed in bacteria did not induce neutralizing antibodies in rabbits. As the majority of the E2 neutralizing epitopes is discontinuous E2 folding in prokaryotes is inaccurate, which demonstrate that a correct conformation of E2 is fundamental for inducing a neutralizing immune response (Yu et al. 1994). The expression of E2 in mammal cells by a recombinant Vaccinia system induced low neutralizing antibody titres in rabbits and bovines (Tijssen et al. 1996).

The E2 protein with the precedent hydrophobic sequence (corresponding to the C- terminal region of the E1 protein) and without its membrane-anchoring region (last 31 aminoacids) expressed in insect cells produced neutralizing antibodies in bovines and rabbits (Bolin and Ridpath 1996; Tijssen et al. 1996). The reported yields were 10 μg E2 ml^{-1} in insect cell lysates infected with recombinant baculovirus (Grigera et al. 2006; Marzocca et al. 2007), and 1–2 μg E2 ml^{-1} in *Escherichia coli* (Toth et al. 1999).

The E2 protein has also been expressed in mammal recombinant viral systems (HEK293T cells) (Donofrio et al. 2006) and has been used as attenuated experimental vaccines (Baxi et al. 2000; Elahi et al. 1999; Kweon et al. 1999).

A chimeral recombinant E2 protein (rE2) was expressed in the transient system Vaccinia and baculovirus (Reed et al. 1999; Grigera et al. 2000; Marzocca 2003). The rE2 has demonstrated identical characteristics to its wild viral chimeral counterpart in terms of its ability to react with antibodies depending in conformation, glycosylation capacity and capacity to form dimmers. Also, rE2 resulted in a potent immunogen in mice, rabbits and bovines being able to induce an anti-BVDV response compatible with the protection of bovines.

Promising results were obtained by the expression of a recombinant glycoprotein E2 in plants (Nelson et al. 2012; Aguirreburualde et al. 2013).

8.7 Plant Systems for Producing Vaccines

The use of plants as a vaccine production platform requires a careful selection of the antigens to be expressed and the development of constructs for a high-level expression of antigens in the plant tissue (Tiwari et al. 2009). Table 8.1 show a brief list of immunogenic antigens expressed in plants. Some of them have demonstrated its preventive and therapeutic value and are in clinical evaluation (Arntzen et al. 2005; Koprowski 2005; Ma et al. 2005; Desgranges 2004; Mason et al. 2002). The classical example is CaroX ® against *Streptococcus mutans* to prevent dental caries (Gavilondo and Larrick 2000; Larrick and Thomas 2001).

Table 8.1 Some examples of immunogenic antigen produced in plants

Disease	Antigen	Species	Reference
Rinderpest	Hemaglutinine	*Arachis hypogaea*	Kandelwal et al. (2004)
Foot and mouth disease	VP1 P1-2A/3C	*Arabidopsis thaliana, Medicago sativa, Solanum tuberosum*	Dus Santos and Wigdorovitz (2005)
Classic Swine Fever	E2	*Medicago sativa*	Legocki et al. (2005)
Rotavirus	VP6	*N. tabacum*	Dong et al. (2005)
Toxoplasmosis	SAG1	*N. tabacum*	Laguía-Becher et al. (2010)
Malaria	Pfs25	*N. benthamiana*	Farrance et al. (2011)

8.8 Expression of the Glycoprotein E2 in *N. tabacum*

The immunogenicity of the recombinant E2 glycoprotein (rE2) expressed in tobacco and its potential use as immunogen in an alternative vaccine was reported (Nelson et al. 2012). Besides the gene for the E2 glycoprotein without its transmembrane domain, the construct carried the signal sequence of the 2S2 seed storage protein of *Arabidopsis thaliana* that drives the protein to the secretory pathway, the KDEL ER–retention signal, and a hexa-histidine tag (His-tag) placed before the KDEL to facilitate protein purification. The derived plasmid pK-2S2-tE2-His-KDEL was introduced in *A. tumefaciens* by electroporation and the resultant *Agrobacterium* was infiltrated in *N. tabacum* leaves of 2-day old plants as was described in (Nelson et al. 2012) The maximum tE2 expression level was attained at the 4th day post-infection. SDS-Page and non- reducing Western blot showed the expression of an E2 dimmer of approximately 80 kDa, like the positive control (Fig. 8.1). The reducing Western blot showed a band of 35 kDa, corresponding to the tE2 monomer.

The expression level was of approximately 20 μg E2 per gram of infected leaf (1.3 % TSP) (Fig. 8.2).

8.8.1 Induction of Specific Neutralizing Antibodies

To analyse the ability of the recombinant E2 expressed in *N. tabacum* to induce specific neutralizing antibodies two different vaccine formulations were tested in guinea pigs. Guinea pigs could replace bovines when testing the immunogenicity of commercial or experimental BVDV vaccines for the optimal correlation between bovine-guinea pig models (Fernández et al. 2009; Jordão et al. 2011). The formulations tested were made with the crude plant extract containing 20 μg ml^{-1} of

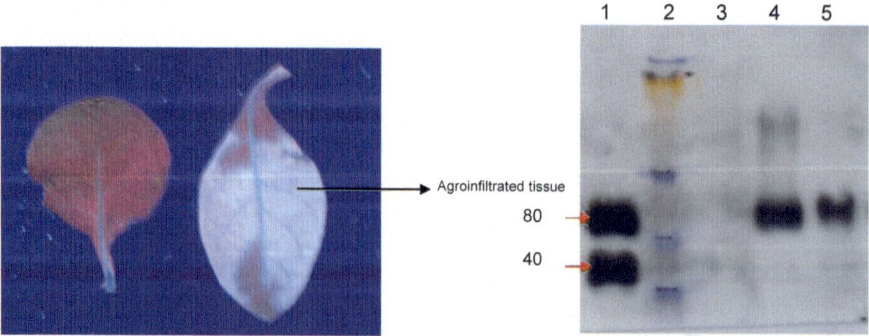

Fig. 8.1 (**a**) *Left*: leaf of *N. tabacum* without treatment, *right*: leaf of agroinfiltrated *N. tabacum* (**b**) Western blot

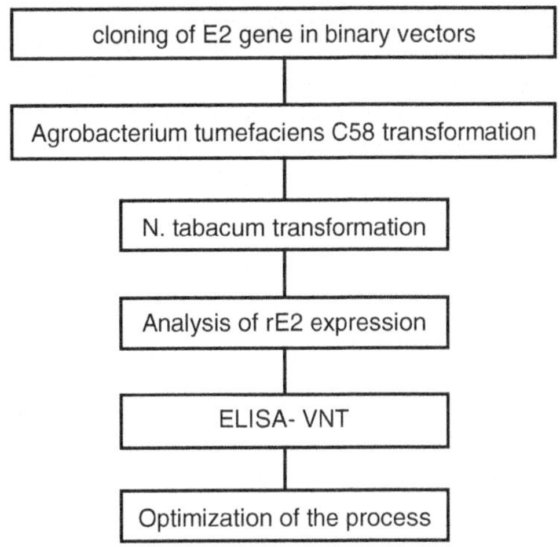

Fig. 8.2 General scheme of the strategy employed for analysing the expression of the recombinant E2 glicoprotein of BVDV in *N. tabacum* and its ability to induce specific neutralizing antibodies

recombinant E2 in a ratio 40:60 with and aqueous adjuvant (Al(OH)3 Hydrogel®) and in a ratio 90:10 with an oily adjuvant (Montanide ISA70 SEPPIC®). Blood samples were taken at time 0 and 15 and 30 days after vaccination. The presence of specific anti-E2 antibodies in blood of vaccinated animals were detected by ELISA and a virus neutralization test (VNT). Values higher than 0.6 corresponding units (OD_{405} or VNT titre) were considered positive.

8.8.2 Evaluation of Anti-E2 Specific Antibodies by ELISA

The titre of the serum was calculated as the reciprocal of the dilution in which OD corresponds to twice the value of the OD of the negative control. Seroconversion was detected in 86 % of animals with the formulation plant extract: oily adjuvant, while it was 75 % with the aqueous adjuvant (Nelson et al. 2012).

8.8.3 Virus Neutralization Test (VNT)

The OIE (2004) recommended the VNT, defined as the loss of infectivity through reaction of the virus with specific antibody, to detect sera neutralizing antibodies (Reed and Muench 1938). The results obtained showed that with the formulation containing the oily adjuvant three out of seven animals produced neutralization antibodies, and that all the animals showed neutralizing antibodies with the aqueous adjuvant. Sham vaccines did not induce neutralizing antibodies (Nelson et al. 2012).

8.9 Conclusive Remarks

Several difficulties arise from the use of the some of the vaccines against VBDV actually in use. Those difficulties are mainly related to the occurrence of secondary effects such as immunosuppression and congenital defects, the emergence of persistent infections by non-cytopathic BVDV biotypes, the in utero infections and/or immunosuppression produced by modified live vaccines, and the introduction of contaminant pathogens by modified-live, attenuated or inactivated vaccines produced in bovine cell cultures (Erickson et al. 1991; Nobiro et al. 2001).

The expression of the recombinant E2 in *N. tabacum*, which is recognized by a specific monoclonal antibody (2.9H), has the ability to generate specific neutralizing antibodies in guinea pigs. These results and those obtained by Aguirreburualde et al. (2013) substantiate more research to establish an alternative plant-based production platform of vaccines against BVDV.

References

Aguirreburualde M, Gómez MC, Ostachuk A, Wolman F, Albanesi G et al (2013) Efficacy of a BVDV subunit vaccine produced in alfalfa transgenic plants. Vet Immunol Immunopathol 151:315–324

Alvarez MA (2012) Is the time coming for plant-made vaccines? J Immunol Techn Infect Dis 1:1–2

Arntzen C, Plotkin S, Dodet B (2005) Plant-derived vaccines and antibodies: potential and limitations. Vaccine 23(15):1753–1756

Aucouturier J, Dupuis L, Ganne V (2001) Adjuvants designed for veterinary and human vaccines. Vaccine 19:2666–2672

Baxi MK, Deregt D, Robertson J, Babiuk LA, Schlapp TY, Tikoo SK (2000) Recombinant bovine adenovirus type 3 expressing bovine viral diarrhea virus glycoprotein E2 induces an immune response in cotton rats. Virology 278:234–243

Bolin SR, Ridpath JF (1995) Assessment of protection from systemic infection or disease afforded by low to intermediate titers of passively acquired neutralizing antibody against bovine viral diarrhea virus in calves. Am J Vet Res 56(6):755–759

Bolin SR, Ridpath JF (1996) Glycoprotein E2 of bovine viral diarrhea virus expressed in insect cells provides calves protected from systemic infection and disease. Arch Virol 141: 1463–1477

Bolin S, Moennig V, Kelso Gourley NEY, Ridpath J (1988) Monoclonal antibodies with neutralizing activity segregate isolate of bovine viral diarrhea virus into groups. Arch Virol 99:117–123

Bracamonte M, Obando C, Plaza N (2006) Diarrea viral bovina. Cómo afecta a los animales. INIA Divulg 8:19–22

Branza-Nichita N, Durantel D, Carrouee-Durantel S, Dwek RA, Zitzmann N (2001) Antiviral effect of N-butyldeoxynojirimycin against bovine viral diarrhea virus correlates with misfolding of E2 envelope proteins and impairment of their association into E1-E2 heterodimers. J Virol 75(8):3527–3536

Ciulli S, Galletti E, Mattilani M, Galligione V, Prosperi S (2009) Analysis of variability and antigenic peptide prediction of E2 BVDV glycoprotein in a mucosal-disease affected animal. Vet Res Commun 33(suppl 1):S125–S127

Corapi WV, Donis RO, Dubovi EJ (1988) Monoclonal antibody analyses of cytopathic and noncytopathic viruses from fatal bovine viral diarrhea virus infections. J Virol 62(8):2823–2827

Deregt D, Bolin SR, van den Hurk J, Ridpath JFY, Gilbert SA (1998a) Mapping of a type 1-specific and type-common epitope on the E2 (gp53) protein of bovine viral diarrhea virus with neutralization escape mutants. Virus Res 53(1):81–90

Deregt D, van Rijn PA, Wiens TYY, van den Hurk J (1998b) Monoclonal antibodies to the E2 protein of a new genotype (type 2) of bovine viral diarrhea virus defines three antigenic domains involved in neutralization. Virus Res 57(2):171–181

Desgranges C (2004) Monoclonal antibodies and therapeutics. Pathol Biol 52(6):351–364

Dong JL, Zhou B, Sheng G, Wang T (2005) Transgenic tobacco expressing a modified VP6 gene protects mice against rotavirus infection. J Integr Plant Biol 47(8):978–987

Donis RO, Corapi W, Dubovi EJ (1988) Neutralizing monoclonal antibodies to BVDV bind to the 56K to 58K glycoprotein. J Gen Virol 69(1):77–86

Donofrio G, Bottarelli E, Sandro C, Flammini C (2006) Expression of bovine diarrhea virus glycoprotein E2 as a soluble secreted form in a mammalian cell line. Clin Vaccine Immunol 13:698–701

Durantel D, Branza-Nichita N, Carrouee-Durantel S, Butters TD, Dwek RA, Yzitzmann N (2001) Study of the mechanism of antiviral action of iminosugar derivatives against bovine viral diarrhea virus. J Virol 75(19):8987–8998

Dus Santos MJ, Wigdorovitz A (2005) Transgenic plants for the production of veterinary vaccines. Immunol Cell Biol 83:229–238

El Omari K, Iourin O, Harlos K, Grimes JM, Stuart D (2013) Structure of a pestivirus envelope glycoprotein E2 clarifies its role in cell entry. Cell Rep 3(1):30–35

Elahi SM, Shen SH, Talbot BG, Massie B, Harpin SY, Elazhary Y (1999) Induction of humoral and cellular responses against the nucleocapsid of bovine viral diarrhea virus by an adenovirus vector with an inducible promoter. Virology 261(1):1–7

Erickson GA, Bolin SR, Landgraf JG (1991) Viral contamination of foetal bovine serum used for tissue culture: risks and concerns. Dev Biol Stand 75:173–175

Farrance CE, Chichester JA, Musiychuk K et al (2011) Antibodies to plant-produced *Plasmodium falciparum* sexual stage protein Pfs25 exhibit transmission blocking activity. Hum Vaccin 7:191–198

Fernández F, Constantini V, Barrandeguy M, Parreno V, Schiappacassp G, Maliandi F (2009) Evaluation of experimental vaccines for bovine viral diarrhea in bovines, ovines and guinea pigs. Rev Argent Microbiol 41:86–91

Fulton RW, Burge LJ (2001) Bovine viral diarrhea virus types 1 and 2 antibody response in calves receiving modified live virus or inactivated vaccines. Vaccine 19(2–3):264–274

Gavilondo JV, Larrick JW (2000) Antibody engineering at the millennium. Biotechniques 29(1): 128–132

Greiser-Wilke I, Dittmar KE, Liess BY, Moennig V (1991) Immunofluorescence studies of biotype-specific expression of bovine viral diarrhea virus epitopes in infected cells. J Gen Virol 72(Pt 8):2015–2019

Grigera PR, Marzocca MP, Capozzo AVE, Buonocuore L, Donis ROY, Rose JK (2000) Presence of bovine viral diarrhea virus (BVDV) E2 glycoprotein in VSV recombinant particles and induction of neutralizing BVDV antibodies in mice. Virus Res 69:3–15

Grigera PR, Marzocca MP, Capozzo A, Buonocore L, Donis R, Rose JK (2006) Presence of bovine viral diarrhea virus (BVDV) E2 glycoprotein in VSV recombinant particles and induction of neutralizing BVDV antibodies in mice. Virus Res 9:3–15

Grummer B, Beer M, Liebler-Tenorio E, Greiser-Wilke I (2001) Localization of viral proteins in cells infected with bovine viral diarrhea virus. J Gen Virol 82(Pt 11):2597–2605

Harkness JW (1987) The control of bovine viral diarrhea virus infection. Ann Rech Vet 18: 167–174

Hefferon KL (2013) Applications of plant-derived vaccines for developing countries. Trop Med Surg 1:1. http://dx.doi.org/10.4172/2329-9088.1000106

Jordão RS, Ribeiro CP, Pituco EM, Okuda LH, Delfava C, de Stefano E et al (2011) Serological response of guinea pigs to oily and aqueous inactivated vaccines containing a Brazilian isolate of the bovine viral diarrhea virus (BVDV). Res Vet Sci. http://dx.doi.org/10.1016/jrvsc.2010.12.010

Kandelwal A, Renukaradhya GJ, Rajeskhar M, Sita GL, Shaila MS (2004) Systemic and oral immunogenicity of hemagglutinin protein of rinderpest virus expressed by transgenic peanut plants in a mouse model. Virology 323(2):284–291

Koprowski H (2005) Vaccines and sera through plant biotechnology. Vaccine 23(15):1757–1763

Kweon CH, Kang SW, Choi EJY, Kang YB (1999) Bovine herpes virus expressing envelope protein (E2) of bovine viral diarrhea virus as a vaccine candidate. J Vet Med Sci 61(4):395–401

Laguía-Becher M, Martín V, Kraemer M et al (2010) Effect of codon optimization and subcellular targeting on *Toxoplasma gondii* antigen SAG1 expression in tobacco leaves to use in subcutaneous and oral immunization in mice. BMC Biotechnol 10:52

Lambrecht B, Kool M, Willart M, Hammad H (2009) Mechanism of action of clinically approved adjuvants. Curr Opin Immunol 21:23–29

Larrick JW, Thomas DW (2001) Producing proteins in transgenic plants and animals. Curr Opin Biotechnol 12:411–418

Legocki A, Miedzinska K, Czaplinska M, Plucieniczak A, Wedrychowicz H (2005) Immunoprotective properties of transgenic plants expressing E2 glycoprotein from CSFV and cysteine protease from *Fasciola hepatica*. Vaccine 23:1844–1846

Lértora WJ (2003) Diarrea viral Bovina: actualización. Rev Vet 14:1–11

Li Y, Wang J, Kanai R, Modis Y (2013) Crystal structure of glycoprotein E2 from bovine viral diarrhea virus. Proc Natl Acad Sci U S A 110(17):6805–6810

Ma J, Barros KC, Bock R, Christou P, Dale P, Dix P, Fischer R et al (2005) Molecular farming for new drugs and vaccines. EMBO Rep 6(7):593–599

Magar R, Minocha HC, Lecomte J (1988) Bovine viral diarrhea virus proteins: heterogeneity of cytopathogenic and non-cytopathogenic strains and evidence of a 53K glycoprotein neutralization epitope. Vet Microbiol 16(4):303–314

Marzocca MP (2003) Caracterización de antígenos recombinantes de la glicoproteína E2 del virus de la Diarrea Viral Bovina y su utilización como vacunas y reactivos de diagnóstico. PhD thesis, Universidad de Buenos Aires, Argentina

Marzocca MP, Seki C, Giambiagi SM, Robiolo B, Schauer R, Dus Santos MJ et al (2007) Truncated E2 of bovine viral diarrhea virus (BVDV) expressed in *Drosophila melanogaster* cells: a candidate antigen for a BVDV ELISA. J Virol Methods 144:49–56

Mason HS, Warzecha H, Mor T, Arntzen CJ (2002) Edible plant vaccines: applications for prophylactic and therapeutic molecular medicine. Trends Mol Med 8(7):324–329

Moennig V, Mateo A, Greiser-Wilke U, Bolin SR, Kelso NE, Liess B (1989) In: Proceedings, abstracts of the annual meeting of the American society for microbiology, Miami Beach, Florida, 8–13 May

Nelson G, Marconi P, Periolo O, La Torre J, Alvarez MA (2012) Immunocompetent truncated E2 glycoprotein of Bovine Viral Diarrhea Virus (BVDV) expressed in *Nicotiana tabacum* plants: a candidate antigen for new generation of veterinary vaccines. Vaccine 30:4499–4504

Nobiro I, Thompson I, Brownlie J, Collins ME (2001) Cytokine adjuvancy of BVDV DNA vaccine enhance both humoral and cellular immune responses in mice. Vaccine 19:4226–4235

O'Hagan, Derek T(2007) New Generation Vaccine Adjuvants. In: eLS. John Wiley & Sons Ltd, Chichester. http://www.els.net [doi: 10.1002/9780470015902.a0020177]

OIE (2004) Manual of standards for diagnostic test and vaccines, 4th ed. p 475

Paton DJ, Lowings JP, Barret AD (1992) Epitope mapping of the gp53 envelope protein of bovine viral diarrhea virus. Virology 190:763–772

Pringle CR (1999) Virus taxonomy at the XIth international congress of virology, Sydney, Australia, 1999. Arch Virol 144(10):2065–2070

Reed LJ, Muench H (1938) A simple method of estimating fifty percent endpoints. Am J Hyg 27:493–497

Reed T, Nettleton PF, Mccrae MA (1999) Expression of the E2 envelope glycoprotein of bovine viral diarrhea virus (BVDV) elicits virus-type specific neutralizing antibodies. Vet Microbiol 65:87–101

Reed SG, Bertholet S, Coler RN, Friede M (2008) New horizons in adjuvants for vaccine development. Trends Immunol 30:23–32

Smith J, Lipstich M, Almond JW (2011) Vaccine production, distribution, access, and uptake. Lancet 378:428–438

Thiel HJ, Stark R, Weiland E, Rumenapf T, Meyers G (1991) Hog cholera virus: molecular composition of virions from a pestivirus. J Virol 65(9):4705–4712

Tijssen P, Pellerin C, Lecomte J, van den Hurk J (1996) Immunodominant E2 (gp 53) sequences of highly virulent bovine viral diarrhea group II viruses indicate a close resemblance to a subgroup of border disease viruses. Virology 217:356–361

Tiwari S, Verma PC, Singh PK, Tuli R (2009) Plant as bioreactors for the production of vaccine antigens. Biotechnol Adv 27:449–467

Toth RL, Nettleton PF, McCrae MA (1999) Expression of the E2 enevelope glycoprotein of bovine viral diarrhoea virus (BVDV) elicits virus-type specific neutralising antibodies. Vet Microbiol 65 (2):87–101

van Oirschot JT, Bruschke CJ, van Rijn PA (1999) Vaccination of cattle against bovine viral diarrhea. Vet Microbiol 64(2–3):169–183

van Rijn PA, van Gennip HG, de Meijer EJ, Moormann RJ (1993) Epitope mapping of envelope glycoprotein E1 of hog cholera virus strain Brescia. J Gen Virol 74(Pt 10):2053–2560

Weiland F, Weiland E, Unger G, Saalmuller A, Thiel HJ (1999) Localization of pestiviral envelope proteins E(rns) and E2 at the cell surface and on isolated particles. J Gen Virol 80:1157–1165

Wensvoort G, Terpstra C (1988) Bovine viral diarrhea virus infections in piglets born to sows vaccinated against swine fever with contaminated vaccine. Res Vet Sci 45(2):143–148

World Health Organization (2013) Guidelines on the nonclinical evaluation of vaccine adjuvants and adjuvanted vaccines

Xu J, Dolan MC, Medrano G, Cramer CL, Weathers PJ (2012) Green factory: plants as bioproduction platforms for recombinant proteins. Biotechnol Adv 30:1171–1184

Xue W, Blecha F, Minocha HC (1990) Antigenic variations in bovine viral diarrhea viruses detected by monoclonal antibodies. J Clin Microbiol 28(8):1688–1693

Yu M, Gould AR, Morrissy CJ, Westbury H (1994) A high level expression of the envelope glycoprotein (gp53) of bovine viral diarrhea virus (Singer) and its potential use as diagnostic reagent. Virus Res 34:178–186

Yu M, Wang LF, Shiell BJ, Morrissy CJ, Westbury HA (1996) Fine mapping of a C-terminal linear epitope highly conserved among the major envelope glycoprotein E2 (gp51 to gp54) of different pestiviruses. Virology 222(1):289–292

Chapter 9
Mathematical Modelling in Recombinant Plant Systems: The Challenge to Produce Heterologous Proteins Under GLP/GMP

Abstract In the last time and in the near future, the markets are demanding large recombinant proteins production capacity, due to the production levels of traditional platforms will be certainly not suffice to make available these demands (Huang and McDonald, Biotechnol Adv 30:398–409, 2012). The explore of alternative tools to assure adequate supplies for a growing demand constitutes a challenge of present times being the "bioengineered pharmaceutical plant" a promising strategy to respond to this require at affordable costs. However, as a commercial process requires high productivities and product yield at minimum cost, and, also, regulations must be follows like GMP and GLP. A mathematical model like the ones described in the present chapter allows making predictions and control of biosynthesis and yielding into the process.

Keywords Mathematical modeling • Recombinant proteins-*Nicotiana tabacum-molecular* farming-cell suspension cultures

9.1 Introduction

According to FDA/USDA (2002) "bioengineered pharmaceutical plant" is a plant manipulated by recombinant DNA technology to express a gene encoding a biological or drug product. Plants expressing transiently o constituently recombinant proteins are considered competitive systems for producing high-value recombinant proteins for medical and industrial purposes. This technology (molecular pharming® or biopharming or agropharming) has advanced greatly as a secure technology, capable of rendering valuable recombinant proteins free of toxins and animal pathogens in a relatively short time and cost-effective (Ma et al. 2005; Franconi et al. 2010; De Muynck et al. 2010; Paul and Ma. 2011; Yuan et al. 2011). This scenario opens up promising horizons to produce recombinant proteins with attractive prospects for commercial exploitation. Many proofs of principles studies just to commercialization have been obtained in the last 25 years (Twyman et al. 2003; Arcalis et al. 2013). Sigma-Aldrich commercializes plant-made recombinant *avidin* and bovine *trypsin*, Dow AgroSciences LLC produces plant-made recombinant veterinary vaccines and it has recently been approved by FDA the enzyme Taliglucerase alfa, a carrot-expressed recombinant glucocerebrosidase for the

M.A. Alvarez, *Plant Biotechnology for Health: From Secondary Metabolites to Molecular Farming*, DOI 10.1007/978-3-319-05771-2_9

treatment of Gaucher disease, to be commercialized by Pfizer (FDA application No (NDA) 022458, 2012). Furthermore, Heber Biotec (Havana, Cuba) has produced the first whole antibody in tobacco plants. It was also demonstrated that most of the recombinant antibodies produced in plants maintains the functional properties as the ones produced in mammalian cell cultures (Kaiser 2008; Zimran et al. 2011; Zeitlin et al. 2011).

9.2 Industrial Production

Manufacturing bioengineered pharmaceutical plants are documented clearly in standard operating procedures (SOPs), Outlines of Production, or other records, following strict regulations (FDA and USDA Regulations 1999, 2002).

Facilities and procedures used for the manufacturing of regulated products, like bioengineered pharmaceutical plants, should be designed following good laboratory practice and good manufacturing processes (GLP and GMP resp.), conditions to ensure pure, sterile, non-toxic, in the case of vaccines assure potency, among other characteristics (Shepherd 1999; EudraLex 2011, ISO 14644-1, Fischer et al. 2012). Product characterization, the manufacturing process, and *in vitro* studies followed by animal studies, are prerequisites to initiating clinical trials. GLP are production and testing practices that helps to ensure a quality product. GLP arranges with the organization, process and conditions under which laboratory studies are planned, performed, monitored, recorded and reported. Also, GLP practices involve a set of regulations designed to establish standards for the conduct and reporting of non-clinical laboratory studies and to ensure the quality and integrity of nonclinical safety data submitted to the regulatory authorities. In that way, GLP data are intended to promote the quality and validity of test data.

GMP has been legislated to pharmaceutical and medical device companies that must follow these procedures, assuring high quality products and batch-to-batch reproducibility. GMPs regulate the manufacture, control, accountability, and documentation of drug or biologic substances and products manufactured for human or veterinary use. The regulations, also, require adequate controls for the release and security of the products before they are allowed to arrive into the general marketplace.

9.3 Industrial Production in Plant Platforms

Plants offer numerous advantages for the production of biopharmaceutical proteins including the capacity to express complex heterologous proteins with posttranslational modifications as glycan patterns and proline hydroxylation among others (Vézina et al. 2009; Eskelin et al. 2009; Gomord et al. 2010; Ahmad et al. 2010;

Xu et al. 2011; Fischer et al. 2012; Strasser 2012). Also, the plant platform has demonstrated to be a secure technology, capable of rendering valuable recombinant proteins free of toxins like retrovirus, prions, mycotoxines and animal pathogens (inherent safety reflecting the inability of human pathogens to replicate in plants) in a relatively short time (Ma et al. 2005; Franconi et al. 2010; De Muynck et al. 2010; Paul and Ma 2011; Yuan et al 2011). Moreover, transient expression has the advantage of a rapid scaled up, therefore offering a rapid response to emerging pandemics or bioterrorism response or the possibility to develop personalized (patient-specific) vaccines (Giritch et al. 2006; Daniell et al. 2009; D'Aoust et al. 2010; Rybicki 2010; Twyman et al. 2012). McCormick et al. (2012) described a plant viral expression system to produce personalized recombinant idiotype vaccines following the concept of a tailor- made programmed. The vaccines against follicular B cell lymphoma were derived from each patient's tumor and it was reported in a phase I clinical trials. Transient tobacco-viral expression system was used for rapid production of the needed amount of idiotype vaccine.

Additionally, other advantages are the low cost of upstream production and the potential for large-scale cultivation (Molina et al. 2004; Escribano and Perez-Filgueira 2009; Thomas et al. 2011; Fischer et al. 2013). The industrial recombinant protein production is achieved by two main platforms: transient expression in plants and fermentation-based platforms with stable expression or transgenic expression in cell suspension cultures.

9.4 Transient Expression Platform

In the last years, bioprocess with binary or viral vectors have also been interested for the rapid and systemic infection and the large amounts of product obtained to achieve an industrial scale. Agro-infiltration, or transient expression system, is a biotechnology method to transiently express a gene during a short period of time without genetic modification (epigenetic modification) in a protoplast, cell, tissue, organ or plant infected. Binary or viral vectors harboring a sequence of the recombinant protein could be introduced in the nucleus of plant target cells by Agrobacterium sp. infection or biobalistic among other technologies (Gleba et al. 2007; Lico et al. 2008; Sainsbury and Lomonossoff 2008; Paul and Ma 2011). The epigenetic expression of the recombinant protein is obtained few days post infection (usually 4–7 days with binary vector or 10–14 days from viral vectors), in significant quantities. After this period of time, the heterologous expression down due to silencing RNAs that encoded to the recombinant protein. Double-stranded RNA is formed between heterologous RNA and complementary endogenous small interfering RNA (siRNA or miRNA). The double strand structure obtained (dsRNA) can trigger posttranscriptional gene silencing (PTGS). The dsRNA activated was cleaved into 22–25 nt RNAs which act as guides to target homologous mRNA sequences for their destruction (Mishra and Mukherjee 2007; Chau and Lee 2007).

For that reason, during transient expression assays never occurred an integration of the transgene into the host genome due to the completely destruction of foreign RNA expression. It is important to point out that transient expression is not considered to be an alteration of genomic plant material and, in consequence, the plant is not considered like GMO (Genetically Modified Organism). Also, cell, tissue, organ or plants transiently transformed do not exhibit macroscopic responses to *A. tumefaciens* infections or bombardment by the gene gun.

The agroinfiltration assays are commonly carried out using *N. tabacum* (tobacco) plants grown in pots and incubated in room chambers. The main reason of use tobacco plants is due to their leaves that are easy to infiltrate and manipulate. Furthermore, tobacco plants are recommended due to the protection against food-chain contamination. Other species agroinfiltrated are tomato and lettuce (Wroblewski et al. 2005).

Also, a variety of strategies were emerging in order to increase yield and achieve an efficient transient platform production during the last decade. One of these emerging technologies is "Magnifection" described as scalable process that can be done on an industrial scale (Gleba et al. 2005). This method involves an efficient vector design carrying T-DNAs encoding RNA replicons combined with and efficient systemic DNA delivery of the viral expression platform by mixing different Agrobacterium lines harbouring fractions of the viral machinery and agrobacteria delivery. In addition, speed, expression level, and yield of a plant RNA virus were drastically increased by altering the codon usage of the virus and the inclusion of typical eukaryotic introns. These modifications were able to the efficiency of delivery, resulting in a reduction in the amount of required infectious agrobacterium (Marillonnet et al. 2005).

In 2006, Bayer and Icon Genetics adopted this methodology and have developed a new production process combining these technologies and transient transformation platform in tobacco plants. Finally, in 2009, Bayer Innovation GmbH present the whole process (http://www.youtube.com/watch?v=fVOBEk5DVZc) following the concept of tailor made pharma.

The FAST technique (Fast Agro-mediated Seedling Transformation) was developed in order to express a wide variety of constructs driven by different promoters in Arabidopsis cotyledons (Li et al. 2009). This method has an especially high potential, ideal for future automated high-throughput analysis (Kolukisaoglu and Thurow 2010). Giritch et al. (2006) using a plant viral vector obtained high expression of heterologous proteins with potential to scale up the process (see below).

ProficiaTM is described as an efficient method to response quickly to emerging diseases (http://www.medicago.com/English/Technologies/Why-Proficia/default.aspx). This platform is described with the high advantage to produce vaccines or therapeutics antibodies in "unmatched speed" obtaining an end product like a vaccine in less than 3 weeks.

Other commercially approaches are extensively reviewed by Paul and Ma (2011).

9.5 Fermentation-Based Platforms

Fermentation-based platforms are carried out in bioreactor batch cell cultures with industrial capacity resulted in high cell densities and production levels. The recombinant or transgenic plants are obtained by introduction a DNA segment (T-DNA), carrying the recombinant protein, into the nucleus of infected plant cells using *Agrobacterium tumefaciens* as biological delivery system or direct transfer methods like biobalistic, microinjection among others (Alvarez and Marconi 2011). Using either indirect or direct methods, it is possible to introduce foreign DNA into any regenerable plant cell type. Using Agrobacterium technology, the T-DNA plasmid containing the heterologous DNA sequence is hazardous inserted in the plant genome, into a chromosome, by illegitimate recombination or non-homologous end-joining (Somers and Makarevich 2004). The stable transformation using Agrobacterium technologies integrate one copy of T-DNA sequence into the cell genome using a T-vector carrying: the recombinant protein, the selectable and the marker genes. After 48 hs post-infection, the recombinant explants must be isolated from the large excess of untransformed cell population using the selectable marker (like antibiotics, herbicides, among others) allowing the recombinant cells proliferation. Cells that do not have the T-DNA integrated into their genome will die, obtaining uniform (nonchimeric) transformants from vegetatively propagated cells. Cloning transformed cells (calli) are induced from recombinant explants grown in a defined media containing growth regulators.

Transformed cells could be grown as microorganism or animal cell culture, easy to scale up. Plant cell bioreactors are obtained in large scale as a safe, convenient and economical production system for recombination proteins in shorter production cycles. This clean technology, growing transgenic material in a confined environment, has been widely accepted by the public perception due to their biosafety implications.

Geneware® expression technology (www.kbpllc.com) developed by Kentucky BioProcessing, LLC (KBP; Owensboro, KY), is specialized on the expression, extraction, and purification of recombinant proteins from plants at commercial scale level. Sigma-Aldrich Fine Chemicals (SAFC; St. Louis, MI) named in the introduction, produce in plant-based platforms therapeutic protein extraction and purification, bio-conjugates, excipients and adjuvants. Medicago Inc., (Quebec City, Canada) manufacturing capability for plant manipulation, product recovery and purification.

Other companies producing plant-based pharmaceuticals are Protalix (Carmiel, Israel) that utilizes a novel bioreactor for plant cell culture system, ProCellExTM, based on disposable sterile plastic bags; Biolex Therapeutical (Pittsboro, NC) based on the aquatic plant *Lemna minor* (www.biolex.com/lexsystem.htm). PharmaPlanta has a pilot scale facility to produce recombinant subunit vaccines, mAbs and other therapeutic proteins, among other products (Huang and McDonald 2012).

9.6 The Main Problem

Commercialization plant-based expression systems were characterized by the lack of a support for large scale facilities manufacturing these products conform to current Good Laboratory and Manufacturing Practice guidelines. The main obstacles to solve are the lack of reproducibility between batches and the variable expression levels within the batch culture. The variations observed in growth and production is a consequence of plant physiology *per se* with genetic and metabolic regulatory controls characteristic of plant kingdom. In transient expression system, the strict temperature control during the growth of plants at the room chamber and the defined post-infiltration harvest-time and the age of leaves to agroinfiltrate are critical for the reproducibility of the results (Buyel 2013). In batch cultures, the plant cells tend to aggregate, are prone to rapid sedimentation, and are vulnerable to high shear sensitivity. For that reason, the manner in which the bioreactor is operated is critical for the reproducibility of the results (Huang and McDonald 2012; Marconi et al. 2014).

The relationship between relative addition rates of nutrients and relative growth rate varies as an unpredictable function in cells and plants. It is due to plants altering periods of assimilation of nutrients with starving stages. The consequence is a pulse on/off between the high nutrient concentration with high availability and relative growth rate, with starving periods where the growth is arrested (Poorter and Lewis 1986; Cannel and Thornlay 2000). During the assimilation period a relative growth rate could be described as:

$$\boxed{\frac{1}{W}\frac{dW}{dt} = \beta\left(N - N_{min}\right)}$$

Where W is weight, β is the slope of the response and N is the nitrogen into the plant and N_{min} is the nitrogen at minimal concentration when growth is arrested. The N is integral to the leaf proteins, being of photosynthetic machinery (Wright et al. 2004).

Taking into account these variations, in growth, and as a consequence, in production parameters, mathematical models could be applied to predict the growth of plant cells being useful tools for the industry. Many predictive mathematical models describing the dynamics of biomass growth, product biosynthesis rates and nutrient substrate consumption are derived from microbial cultures. Also, the complexity of models varies enormously, being affected by the abilities, available data and the objectives of the modelers. However, it is expected that a mathematical model can be used to predict or represent or to describe the behavior of a system using mathematical language. Also, models must facilitate process optimization, including of increasing product quality and productivity and reducing manufacturing cost, risk, and time. In the present chapter, two models are present for transient and stably plant and cell cultures.

9.7 Mathematical Modeling for Transient Platform

Johannes F. Buyel (2013) presented an interesting Thesis where a mathematical modeling is developed for plant transient expression systems. The protein model to be express was a monoclonal antibody 2G12 described as specific for the HIV gp120 glycoprotein. Also, a marker protein, DsRED was used as tracer in the experiments. The set of experiments was done using transient expression in *Nicotiana tabacum* var. Petit Havana SR1 by infiltration of leaves with *Agrobacterium tumefaciens* harboring the recombinant plasmid.

The infiltrated leaf was subdivided into four positions parallel to the main vein as a matrix (Fig. 9.1). This variable was indicated by the suffix *p*. Also, the plant leaves were identified by age since the number 1 assigned for the oldest to the apical leaf (number 8). The suffix to denote the individual leaf into de plant was *k* (same figure).

The quantity and distribution of the recombinant protein could be processed like a mathematical matrix. The initial concept is described by this equation:

$$C[P] = \frac{m[P]}{V}$$

Where, C[P] is the recombinant protein concentration and m[P] is the mass of the recombinant protein in a known extract volume (V).

The matrix could be described as:

$$V_{k,p} = m_{t,k} \cdot \eta_k \cdot \Psi_{k,p} \cdot \varepsilon$$

Where the extract volume (V) for the position *p* in a *k* leaf has a mass average of m in this leaf *k*, η modified by effective biomass ratio for each *k* leaf (calculated by the intercostal field biomass (g) per biomass extracted (g)) and the position correction Ψ for each leaf *k* (biomass of position *p* per biomass extracted in g). Finally, ε is the

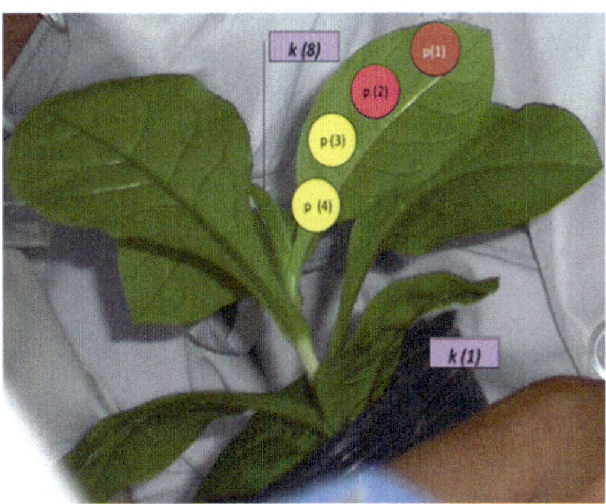

Fig. 9.1 *N. tabacum* leaves to be used to transiently express a protein

extract ratio calculated by the volume of solids-free extract (mL) per biomass extracted (g).

The final recombinant protein yield is obtained from a linear second-order polynomial fit:

$$\int(V_{k,p}, \beta_k) = \beta_{k,1} + \beta_{k,2} V_{k,p} + \beta_{k,3} (V_{k,p})^2; \beta_k = (\beta_{k,1}, \beta_{k,2}, \beta_{k,3})$$

The study includes a quantitative impact of many parameters, determined and modeled using a design of experimental approach. The post-infiltration incubation temperature, plant and leaf age and incubation time with Agrobacterium were found to be major factors influencing protein yields and growth variation.

9.8 Mathematical Modeling for Fermentation-Based Platforms

The recombinant full antibody r14D9 was used as a protein model to be expressed in *Nicotiana tabacum* cv. Xhanti NN cell suspension cultures (Petruccelli et al. 2006; López et al. 2010). For the bioreactor production, cell suspension expressing the antibody r14D9 with the retention signal KDEL at endoplasmic reticulum were scaled up from a 225 mL Erlenmeyer flask to 2L-bioreactor. Growing cell suspension cultures in a batch bioreactor is often complicated due to the cell aggregation process that leads to the formation of zones where the cell growth is limited. The growth could be restricted by oxygen and nutrients demands, increase the heterogeneity of culture and concentration of toxic substances, among others. Also, plants cells have a critical inoculation density below which growth is arrested. The consequence will be the accumulation of non-dividing cells in a short period of time, and finally aborting the experiments.

Typical slope of the growth kinetics and production of recombinant protein have high standard error with high dispersion of scores around the means (Fig. 9.2). This curves could be characterized with an early exponential period with low recombinant protein production, a gradual slowdown as dividing cell reaches its maximum and a subsequent stationary face (same figure). The recombinant protein production shows more unpredictable features. However, the recombinant protein production curve suggests the production rate follows the dividing or proliferating cells concentration.

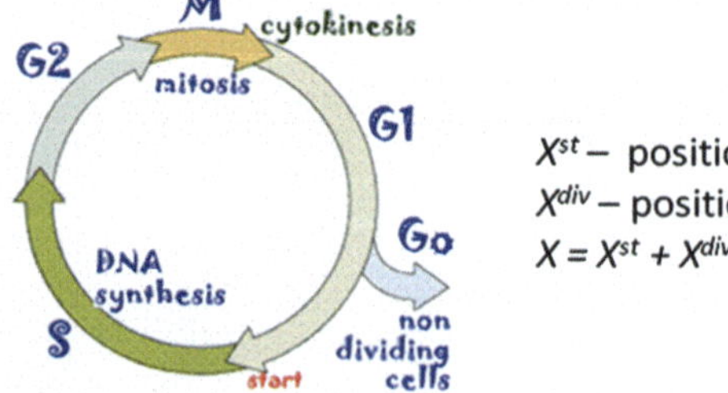

Fig. 9.2 Schematic representation of plant cell cycle

As a result, the cell population rapidly differentiates into dividing (growing and dividing cells) and non-dividing (stable) cells and each population growth with a proper kinetics. Thus, in order to calculate the dynamics of cell growth it is necessary to take into account the difference in energy substrate metabolism of stable and proliferating cells in such zones and use both structured and unstructured models that were previously successfully applied for microbial batch processes (Klykov and Kurakov 2012, 2013).

The mathematical model is based on the theory of energy-limited growth that allows precise evaluation of a cell population age structure into the bioreactor.

The model is based on treating the biomass as two main groups: dividing and non-dividing cells (using a combination of statistical data and qualitative causal assumptions) (Fig. 9.2). Therefore, a phytofermentation process is a "mixture" of cells of different ages being this population characterized by the parameter R:

$$\boxed{R = \frac{X^{st}}{X}}$$

Where X^{st} is the non-dividing cell concentration at the synchronized degree R, and X is the total biomass. The age structure of the cell population varies, thus characterizing these variations is fundamental for the structured model.

On the other hand, the unstructured model is based on the decrease of the absolute and the specific growth rates of the biomass, as these parameters are directly related to the biomass concentration at the growth limitation phase when oxygen is limited (GIP: growth inhibition phase):

$$\boxed{A = \frac{m}{a}}$$

In this model, the energy substrate consumption rate for viability maintenance (m) is specified by the oxygen and trophic coefficient (a) that is assumed as a constant during GIP. The population is synchronized to non-proliferating cells, which consume oxygen only for primary metabolism.

Equation of the structured model for biomass in the GIP is:

$$\boxed{\frac{d^n X^{div}}{d\left(X^{st}\right)^n} = \frac{K(-1)^{(n-1)}\, n!}{A^2\left(X^{st}\right)^{(n+1)}} - C}$$

Where n – is an integer that specifies the order of the derivative of this function; X^{div} – is the quantity of dividing cells; X^{st} – is the quantity of non-dividing (stable) cells; k – is the overall growth rate multiplied by the rate of accumulation of stable (or non-dividing) cells; A – is the ratio of energy required to maintain viability to energy required for biomass accumulation and/or maintaining the rate of accumulation of stable (non-dividing) cells. In addition, if $n=1$ then $C=1$, if $n \geq 2$ then $C=0$.

The equation of a structured model for substrate (metabolites) in growth inhibition phase is:

$$d(P \text{ or } -S)/d\tau = k^{div}{}_{P,S}X^{div} + k^{st}{}_{P,S}X^{st}$$

P and S symbols are the valuable metabolite and the substrate, respectively, $k^{div}{}_{P,S}$ and $k^{st}{}_{P,S}$ coefficients are constants synthesis (utilization) of dividing and non-dividing cells. In this model, it is assumed that metabolites are synthesized only by proliferating (dividing) cells. Non-proliferating (stable) cells, as a rule, destroy these products. Therefore, signs of the constants for metabolite synthesis and degradation are opposite. The same should be stated for substrates utilized for cell construction.

The equation for the unstructured model in growth inhibition phase (GIP) is:

$$X = X_p - (X_p - X_{lim}) * \exp^{(-A*(\tau - \tau\ lim))}$$

Where X is the amount of biomass calculated using an unstructured model; X_p – the maximum estimated amount of biomass; X_{Lim} – the amount of biomass at the start of the limitation of cell growth; τ and τ_{lim} are the terms of the estimated time of cultivation and cultivation duration from start until the beginning of the limitation of cell growth.

These equations describe all the known diversity of the processes with S-like growth curves and changes in the concentrations of substances in closed systems, which is an entirely new and previously unknown fact. It is known that a physical law means a generalization of a numerical relationship between the objects of the real physical world that is running under specified conditions for the class of the objects and does not follow from any of the previously discovered laws. There is no reason not to admit the two described equations for the GIP as laws for GIP. The data obtained were used for the selection of techniques to increase the protein expression by genetically modified microorganisms (Klykov et al. 2011).

The proposed modeling approach for the cultivation of plant cells using a structured model reflects the real dynamics of changes in the age structure of cell populations in the batch culture. Isolating growing cells based on physiological age into two groups – dividing and stable cells (non-dividing), allows applying a structured model for scaling up from the bench to a phytofermenter. Processing data, obtained during cultivation and according to the structured model, gives the opportunity to deeper understand the dynamics of changes occurring in the cell population, accurately predict both the dynamics of biomass growth and biosynthesis of a target protein or metabolite and significantly reduce the cost of improving cultivation of cells on an industrial scale.

9.9 Conclusions

Development of transgenic plants is a technology used to basic studies like functional genomics research and, also, in modern plant breeding and in the last years as a platform for recombinant proteins production. There are many protocols to obtain transient or transgenic plants expressing heterologous proteins. In both systems, heterologous protein production like transient platforms or fermentation-based platforms proved to be a suitable process to obtain large- scale productions. However, a commercial process requires high productivities and product yield at minimum cost (Scragg 1992), and, also, regulations must be follows like GMP and GLP. A mathematical model like the ones described before allows making predictions and control of biosynthesis and yielding into the process.

References

Ahmad A, Pereira EO, Conley AJ, Richman AS, Menassa R (2010) Green biofactories: recombinant protein production in plants. Recent Pat Biotechnol 4:260–262

Alvarez MA, Marconi PL (2011) Genetic transformation for metabolic engineering of tropane alkaloids. In: Alvarez MA (ed) Genetic transformation, vol 15. Intech Publisher, Croatia, pp 291–304

Arcalis E, Stadlmann J, Rademacher T, Marcel S, Sack M, Altmann F, Stoger E (2013) Plant species and organ influence the structure and subcellular localization of recombinant glycoproteins. Plant Mol Biol 83:105–117

Buyel T (2013) Manufacturing biopharmaceutical proteins by transient expression in *Nicotiana tabacum* (L.). PhD thesis, Institute for Molecular Biotechnology at the RWTH Aachen University

Cannel MGR, Thornlay JHM (2000) Modelling the components of plant respiration: some guiding principles. Ann Bot 85:45–54

Chau BL, Lee K (2007) Function and anatomy of plant siRNA pools derived from hairpin transgenes. Plant Methods 3:13–37

D'Aoust MA, Couture MM, Charland N, Trépanier S, Landry N, Ors F et al (2010) The production of hemagglutinin-based virus-like particles in plants: a rapid, efficient and safe response to pandemic influenza. Plant Biotechnol J 8:607–619

Daniell H, Singh ND, Mason H, Streatfield SJ (2009) Plant-made vaccine antigens and biopharmaceuticals. Trends Plant Sc 14:669–679

De Muynck B, Navarre C, Boutry M (2010) Production of antibodies in plants: status after twenty years. J Plant Biotechnol 8:529–563

Escribano JM, Perez-Filgueira DM (2009) Strategies for improving vaccine antigens expression in transgenic plants: fusion to carrier sequences. Methods Mol Biol 483:275–287

Eskelin K, Ritala A, Suntio T, Blumer S, Holkeri H, Wahlstro EH, Baez J, Ma K, Nuutila AM (2009) Production of a recombinant full-length collagen type Ia-1 and of a 45-kDa collagen type I a-1 fragment in barley seeds. Plant Biotechnol J 7:657–672

EudraLex (2011) Volume 4: The rules governing medicinal products in the European Union. Part I: Medicinal products for human and veterinary use. EU guidelines for good manufacturing practice. EU Publisher, Brussels

FDA (1999) Draft guidance for industry: cooperative manufacturing arrangements for licensed biologics. US FDA, Rockville

FDA/USDA (2002) Draft guidance. Drugs, biologics, and medical devices derived from bioengineered plants for use in humans and animals. FDA, Rockville MD. http://www.fda.gov/

downloads/AnimalVeterinary/GuidanceComplianceEnforcement/GuidanceforIndustry/UCM055424.pdf

Fischer R, Schillberg S, Hellwing S, Twyman RM, Drossard J (2012) GMP issues for recombinant plant-derived pharmaceuticals proteins. Biotechnol Adv 30:434–439

Fischer R, Schillberg S, Buyel JF, Twyman RM (2013) Commercial aspects of pharmaceutical protein production in plants. Curr Pharm Des 19:5471–5477

Franconi R, Demurtas OC, Massa S (2010) Plant-derived vaccines and others therapeutics produced in contained systems. Expert Rev Vaccines 9:877–892

Giritch A, Marillonet S, Engler C, van Eldik G, Botterman J, Klimiuk V et al (2006) Rapid high-yield expression of full-size IgG antibodies in plants coinfected with noncompeting viral vectors. Proc Natl Acad Sci USA 103:4701–14706

Gleba Y, Klimyuk V, Marillonnet S (2005) Magnifection – a new platform for expressing recombinant vaccines in plants. Vaccine 23:2042–2048

Gleba Y, Klimyuk V, Marillonnet S (2007) Viral vectors for the expression of proteins in plants. Curr Opin Biotechnol 18:134–141

Gomord G, Fitchette AC, Menu-Bouaouiche L, Saint-Jore-Dupas C, Plasson C, Michaud D et al (2010) Plant-specific glycosylation patterns in the context of therapeutic protein production. Plant Biotechnol J 8:564–587

Huang T-K, McDonald KA (2012) Bioreactor systems for in vitro production of foreign proteins using plant cell cultures. Biotechnol Adv 30:398–409

Kaiser J (2008) Is the drought over for pharming? Science 320:473–475

Klykov SP, Kurakov VV (2012) New kinetic structured model for cell cultivation in a chemostat. BioProcess Int 10:36–43

Klykov SP, Kurakov VV (2013) Analysis of bacterial biomass growth and metabolic accumulation. BioProcess Tech 11:36–43

Klykov SP, Kurakov VV, Vilkov VB, Demidyuk IV, Gromov TY, Skladnev DA (2011) A cell population structuring model to estimate recombinant strain growth in a closed system for subsequent search of the mode to increase protein accumulation during protealysin producer cultivation. Biofabrication 3:045006

Kolukisaoglu Ü, Thurow K (2010) Future and frontiers of automated screening in plant sciences. Plant Sci 178:476–484

Li JF, Park E, von Arnim A, Nebenfuhr A (2009) The FAST technique: a simplified Agrobacterium-based transformation method for transient gene expression analysis in seedlings of Arabidopsis and other plant species. Plant Methods 5:6

Lico C, Chen Q, Santi L (2008) Viral vectors for production of recombinant proteins in plants. J Cell Physiol 216:366–377

López J, Lencina F, Petruccelli S, Marconi PL, Alvarez MA (2010) Influence of the KDEL signal, DMSO and mannitol on the production of the recombinant antibody 14D9 by long-term *Nicotiana tabacum* cell suspension culture. Plant Cell Tissue Organ Cult 103:307–314

Ma J, Barros K-C, Bock E, Dale RCP, Dix PJ, Fischer R, Irwin J et al (2005) Current perspectives on the production of pharmaceuticals in transgenic plants. EMBO Rep 6:593–599

Marconi PL, Alvarez MA, Klykov SP, Kurakov VV (2014) Mathematical model for production of recombinant antibody 14D9 by *Nicotiana tabacum* cell suspension batch culture. BioProcess Int 12:42–49

Marillonnet S, Thoeringer C, Kandzia R, Klimyuk V, Gleba Y (2005) Systemic *Agrobacterium tumefaciens*-mediated transfection of viral replicons for efficient transient expression in plants. Nat Biotechnol 23:718–723

McCormik AA, Reddy S, Reinl SJ, Cameron TI, Czerwinkski DK, Vodjani F et al (2012) Plant-produced idiotype vaccines for the treatment of non-Hodgkin's lymphoma: safety and immunogenicity in a phase I clinical studies. Proc Natl Acad Sci USA 105:10131–10136

Mishra NS, Mukherjee SK (2007) A peep into the plant miRNA world. Open Plant J 1:1–9

Molina A, Hervas-Stubbs S, Daniell H, Mingo-Castel AM, Veramendi J (2004) High-yield expression of a viral peptide animal vaccine in transgenic tobacco chloroplasts. Plant Biotechnol J 2:141–153

Paul C, Ma J (2011) Plant-made pharmaceuticals: leading products and production platforms. Biotechnol Appl Biochem 58:58–67

Petruccelli S, Otegui MS, Lareu F, Trandinhthanhlien O, Fitchette A-C, Circosta A, Rumbo M, Bardor M, Carcamo R, Gomord V, Beachy RN (2006) A KDEL-tagged monoclonal antibody is efficiently retained in the endoplasmic reticulum in leaves, but is both partially secreted and sorted to protein storage vacuoles in seeds. J Plant Biotechnol 4:511–527

Poorter H, Lewis C (1986) Testing differences in relative growth rate: a method avoiding curve fitting and pairing. Physiol Plant 67:223–226

Rybicki EP (2010) Plant-made vaccines for humans and animals. Plant Biotechnol J 8:620–637

Sainsbury F, Lomonossoff G (2008) Extremely high-level and rapid transient protein production in plants without the use of viral replication. Plant Physiol 148:1212–1218

Scragg AH (1992) Chapter 4: Bioreactor for the mass cultivation of cell plants. In: Fowler MW, Waren GS (eds) Plant Biotechnology. Pergamon Press, Oxford

Shepherd AJ (1999) Chapter 19: Good laboratory practice in the research and development laboratory. In: Meager A (ed) Gene therapy technologies, applications and regulations. Wiley, Chichester, pp 375–381

Somers DA, Makarevich I (2004) Transgene integration in plants: poking or patching holes in promiscuous genomes? Curr Opin Biotechnol 15:126–131

Strasser R (2012) Challenges in O-glycan engineering in plants. Front Plant Sci 3:218

Thomas DR, Penney CA, Majumder A, Walmsley AM (2011) Evolution of plant-made pharmaceuticals. Int J Mol Sci 12:3220–3236. Molecular farming in plants: host systems and expression technology. Trends Biotechnol 21:570–578

Twyman RM, Stoger E, Schillberg S, Christou P, Fischer R (2003) Molecular farming in plants: host systems and expression technology. Trends Biotechnol 21:570–578

Vezina LP, Faye L, Lerouge P, D'Aoust MA, Marquet-Blouin E, Burel C, Lavoie PO, Bardor M, Gomord V (2009) Transient co-expression for fast and high-yield production of antibodies with human-like N-glycans in plants. Plant Biotechnol J 7:442–455

Wright IJ, Reich PB, Westoby M, Ackerly D, Zdravko B, Bongers F, Cavender-Bares J, Chapin T, Cornelissen JHC, Diemer M, Flexas J, Garnier E, Groom PK, Gulias J, Hikosaka K, Lamont BB, Lee T et al (2004) The world wild leaf economic spectrum. Nature 428:821–827

Wroblewski T, Tomczak A, Michelmore R (2005) Optimization of Agrobacterium mediated transient assays of gene expression in lettuce, tomato and Arabidopsis. Plant Biotechnol J 3:259–273

Xu J, Ge X, Dolan MC (2011) Towards high-yield production of pharmaceutical proteins with plant cell suspension cultures. Biotechnol Adv 29:278–299

Yuan D, Bassie L, Sabalza M, Miralpeix B, Dashevskaya S, Farre G, Rivera S et al (2011) The potential impact of plant biotechnology on the Millennium Development Goals. Plant Cell Rep 30:249–265

Zeitlin L, Pettitt J, Scully C, Bohorova N, Kim D, Pauly M, Hiatt A, Ngo L, Steinkellner H, Whaley KJ, Olinger GG (2011) Enhanced potency of a fucose-free monoclonal antibody being developed as an Ebola virus immunoprotectant. Proc Natl Acad Sci USA 108:20690–20694

Zimran A, Brill-Almon E, Chertkoff R, Petakov M, Blanco-Favela F, Terreros Muñoz E et al (2011) Pivotal trial with plant cell–expressed recombinant glucocerebrosidase, taliglucerase alfa, a novel enzyme replacement therapy for Gaucher disease. Blood 118:5567–5773

ERRATUM TO

Mathematical Modelling in Recombinant Plant Systems: The Challenge to Produce Heterologous Proteins Under GLP/GMP

María Alejandra Alvarez

M.A. Alvarez, *Plant Biotechnology for Health: From Secondary Metabolites to Molecular Farming*, DOI 10.1007/978-3-319-05771-2_9

DOI 10.1007/978-3-319-05771-2_10

This chapter was contributed by three authors. Inadvertently the chapter was published without the below mentioned authors;

P. L. Marconi (✉)
U. Maimonides-CONICET, Hidalgo 7758, CABA, Argentina
e-mail: marconi.patricialaura@maimonides.edu

S. Klykov (✉)
PHARM-REGION, Ltd, c.1, b.10, Baryshikha street, Moscow, Russia
e-mail: SmlK03@mail.ru

The online version of the original chapter can be found at http://dx.doi.org/10.1007/978-3-319-05771-2_9

M.A. Alvarez, *Plant Biotechnology for Health: From Secondary Metabolites to Molecular Farming*, DOI 10.1007/978-3-319-05771-2_10

Index

M.A. Alvarez, *Plant Biotechnology for Health: From Secondary Metabolites to Molecular Farming*, DOI 10.1007/978-3-319-05771-2

The manufacturer's authorised representative in the EU is Springer Nature Customer Service Centre GmbH, Europaplatz 3, 69115 Heidelberg, Germany. If you have any concerns regarding our products, please contact ProductSafety@springernature.com

Printed and bound by CPI Group (UK) Ltd, Croydon, CR0 4YY
15/07/2026
02167649-0001